HANDBOOK OF ELECTRONICS

HANDBOOK OF ELECTRONICS

V.K. Puri, M.Sc., Ph.D
P. Chand, M.Sc., Ph.D

ANMOL PUBLICATIONS PVT LTD
New Delhi-110 002

ANMOL PUBLICATIONS PVT LTD
4374 / 4B, Ansari Road
Daryaganj, New Delhi-110 002

Handbook of Electronics

ISBN 81-7041-967-0
Reprint : 1996

PRINTED IN INDIA

Published by J.L. Kumar for Anmol Publications Pvt. Ltd., New Delhi and Printed at Mehra Offset Press, Delhi.

Preface

Handbook of Electronics is an attempt to put the important facts and other important terms/definitions/formulae etc. related to various topics in Electronics.

The book has been divided into different chapters to facilitate its use and easy accessibility of the terms, definitions etc. to the reader. Each topic is followed by objective type questions to test the knowledge gained by the learner. These questions have been planned in such a way so as to compel the learner to use his ingenuity and knowledge to arrive at a correct answer. These questions are very important from the examination point of view also.

It is expected that the book will be useful to college students, university students and the students appearing for various competitive examination. It will prove to be a handy book for the quick revision of various topics included in this book.

It is hopefully expected that the book will be liked by teachers and students alike.

Suggestions for improvement of the book shall be thankfully accepted.

Authors

Contents

Chapter 1

Electronics—Fundamentals

STUDY MATERIAL

Basic Particles and Data, about Atom

Mass of electron = 9.11×10^{-31} kg

Mass of proton = 1.67×10^{-27} kg

Mass of neutron $\simeq$ mass of proton

Mass of electron $\simeq \frac{1}{1840}$ mass of proton

Diameter of nucleus is of the order of 10^{-14} m

Diameter of orbits $\simeq 10^{4}$ times dia of molecule.

Diameter of electron $\simeq 10^{-15}$m

Charge on electron = -1.602×10^{-19} coulomb

Charge on proton = $+1.602 \times 10^{19}$ coulomb

When an electron is taken out from a neutron atom, this atom becomes positively charged and is known as a **positive ion,** and if an electron is added to a neural atom, it becomes negatively charged and is known as a **negative ion**. Thus, by gain or loss of an electron an **atom** becomes an **ion.**

A force of field exists between ions. Two ions of similar charge will repel each other whereas ions of opposite charges will attract each other. The field of force between ion is termed as electric, dielectric or electrostatic field. The movement of ions, arising from the presence of this field is called **the electric current.**

The charge on an electron is measured in terms of **coulomb**. The unit of current is coulomb per second is called the **ampere**. Thus

$$\text{Ampere} = \frac{\text{coulomb}}{\text{second}}$$

One coulomb is equivalent to the charge of 6.28×10^{18} electrons.

Electromotive Force or potential of a body is the workdone in joules to bring a unit electric charge from infinity to the body. It is expressed in terms of *volts*.

The **potential, difference** is defined as that which causes current to flow in the **closed circuit**.

Resistance is the property of a substance due to which it opposes the flow of electrons (i.e., electric current) through it. The **unit of resistance** is ohm (Ω).

Metals, acids and salt solutions are good conductors of electricity. Silver, copper and aluminium offer least resistance to flow of current and are called very good conductor of electricity. **The electrons while flowing through the molecules or the atoms of the conductor, collide with other atoms and electrons, thereby producing heat.**

Some substances offer relatively greater difficulty or hindrance to the passage of these electrons. Such substances are called poor conductors or insulators of electricity. Some of the insulators are glass, bakelite, mica, rubber, polyvinyl chloride (P.V.C.), dry wood, etc.

Factors on which the resistance of a conductor depends:

(i) *Length of conductor*—it varies directly with the length;

(ii) *Cross-sectional area of the conductor*—it varies

inersely with the cross-sectional area;

(iii) *Its resistivity i.e.* the nature of composition, etc., of the material of which the conductor is made up; and

(iv) *Temperature of the conductor*— it almost varies directly with the temperature. Thus R, the resistance of a conductor is given by

$$R = \rho \frac{l}{A}$$

where ρ = specific resistance of resistivity of the material.
l = length of the conductors
A = cross-sectional area of conductor.

Ohm's Law

If the temperature and other conditions remain constant, the current through a conductor is proportional to the applied potential difference and it remains constant. Thus

$$\text{Current} = \frac{\text{Applied voltage}}{\text{Resistance of the circuit}}$$

$$\text{Resistance} = \frac{\text{Applied voltage}}{\text{Current in the circuit}}$$

Potential across resistance = Current × Resistance

Conditions for Ohm's Law

(1) Ohm's law can be applied either to the entire circuit or a part of a circuit.

(2) When ohm's law is applied to a part circuit, part resistance and the potential across the part resistance should be used.

(3) The Ohm's law can be applied to d.c. as well as a.c. circuits. However, in case of a.c. circuits impedance Z, is used in place of resistance. Thus

$$I = \frac{E}{Z} = \frac{\text{Applied voltage}}{\text{Impedance in the circuit}}$$

Limitations of Ohm's Law

Ohm's law is not applicable under the following conditions:

1. For metals which get heated up due to flow of current through them.
2. For electrolytes where enormous gases are produced on either electrode.
3. For vacuum radio valves.
4. For gas filled tubes, in which the ions are generated as a result of current flow.
5. For arc lamps.
6. For semiconductors.

Tolerance of an electronic component, whether R, L or C, gives percentage plus and minus variation that may occur in ohms rating, inductance value or capacitance value respectively.

Colour-code for resistors. It is given by the following colours in order from 0 → 9, viz., Black, Brown, Red, Orange, Yellow, Green, Blue, Violet, Grey and White with Black representing zero and while representing 9. Colour code is done by having four strips, three on the left side and one on the right side. Out of the three, first and second give first two digits and third and decimal multiplier. Colour strip on right gives tolerance. Golden represents ± 5% whereas silver represents ± 10%, ± 20% is represented by absence of colour strip.

For resistance connected in series the total resistance is equal to the sum of all resistances. i.e., $R = R_1 + R_2 + \ldots$ In parallel mode, the total resistance R is given by

$$\frac{1}{R} = \frac{1}{R_1} + \frac{1}{R_2} + \frac{1}{R_3} + ...$$

Moderate and high values of resistances are made as carbon resistors while very low value of resistances are achieved with wire-wound type of resistors.

A current carrying conductor is accompanied by a magnetic field surrounding it and generation of heat equal to I^2R.

Inductance of a coil. It is given by $L = \frac{\mu AN^2}{l}$ H, where μ is permeability, A the area of cross-section, N the no. of turns and l the length. When A and l are in m^2, and m then L is in Henrys.

Voltage across an inductance leads the current by 90°. Reactance of an inductor is ($2\,\mu f L$) where f is the frequency.

A transformer. It is a device that transforms electrical energy from circuit to another by magnetic means. In transformers, $\frac{E_2}{E_2} = \frac{I_2}{I_1}$

Reactance of a capacitor. It is given by $\frac{1}{2\pi f C}$, where f is the frequency. Current through a capacitor leads the voltage by 90°.

A trimmer. It is a variable capacitor set for best performance. A gang is a variable air condenser. All high voltage high current discharge capacitors are paper capacitors. Very low values of capacitances are achieved in mica capacitors.

C_1 and C_2 connected in series given C equal to C_1C_2/C_1+C_2 and connected in parallel give ($U_1 + C_2$)

L_1 and L_2 connected in series given L equal to ($L_1 + L_2$) and connected in parallel give $\frac{L_1 L_2}{L_1 + L_2}$.

Xc_1 and Xc_2 in series give $Xc_1 + Xc_2$ and in parallel give $\frac{Xc_1\ Xc_2}{Xc_1 + Xc_2}$. Some is the case with inductive reactances.

X_{L_1} and X_{L_2} in series given $X_{L_1} + X_{L_2}$ and in parallel give $\left(\frac{Xc_1\ Xc_2}{X_{L_1} + X_{L_2}}\right)$

A series resonant circuit has a resonant frequency $f = \frac{1}{2\pi\sqrt{LC}}$. Resonance is said to occur when $X_L = X_C$.

A parallel resonant frequency circuit's resonance frequency $f = \frac{1}{2\pi\sqrt{LC}}$. Parallel resonance occurs when $X_L = X_C$.

Thevenin's theorem. It reduces a complicated network to a voltage generator in series with an impedance.

Notron's theorem. It reduces a complicated network to a current generator in parallel with an impedance.

Maximum power transfer theorem. According to it power transfer is maximum when source impedance is complex conjugate of load impedance.

Capacitor charging. It follows an exponential growth curve.

Capacitor discharge. It follows an exponential decay curve.

Current through an inductor. It follows an exponential growth during charging.

An inductor opposes a change in voltage whereas a capacitor opposes a change in current. Voltage (instantaneous) across a capacitor is given by $v = \frac{1}{C} \int i \, dt$ and current (instantaneous) through an inductor is given by $i = \frac{1}{L} \int v \, dt$.

Current (instantaneous) through a capacitor. It is given by $i = \frac{C\,dv}{dt}$ and voltage across in inductor is given by $v = \frac{L\,di}{dt}$.

Wavelength (λ) of an electronic signal. It given by $\lambda = \frac{C}{f}$, where C is velocity of electromagnetic waves and f is the frequency.

Hall Effect. According to it "when a conductor, with current I passing through its axis, is placed in a magnetic field, an EMF appears across its end surfaces."

Lowering of work function of an emitting surface by applying an electric field along its surface is known as **Schottky Effect.**

OBJECTIVE TYPE QUESTIONS

Q. 1. What is the impedance across terminals A-A in the figure 1.1?

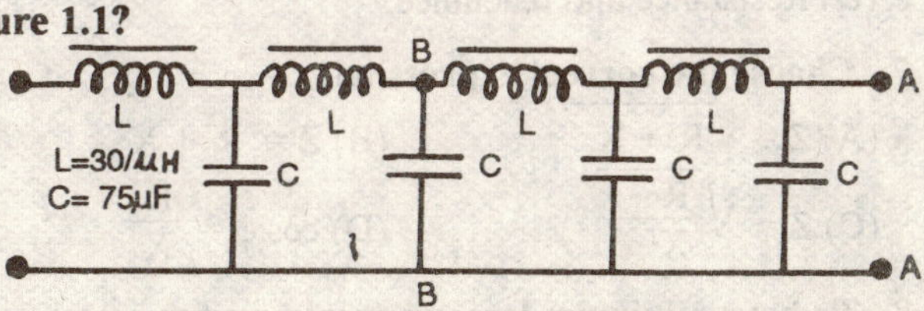

Fig. 1.1

(A) 2 Ω (B) 4 Ω
(C) Indeterminate (D) None of these.

Q. 2. A 220 V, 50 Hz AC supply signifies that the RMS value

of voltage waveform is

(A) 220 V (B) 220 $\sqrt{2}$ V.

(C) 220/ $\sqrt{2}$ V (D) All of these.

Q. 3. Form factor of a 220 V, 50 Hz AC waveform is

(A) 1.414 (B) 1.50

(C) 4.50 (D) None of these.

Q. 4. A capacitor C = 10 μF is charged from a d.c. source of 100 V. When discharged into a load of 2 Ω, it gives a narrow pulse of duration

(A) 80 μ sec. (B) 30 μ sec.

(C) 0.055 μ sec. (D) None of these.

Q. 5. If voltage and current are 90° out of phase, power consumed will be

(A) Maximum (B) Minimum

(C) Zero (D) None of these.

Q. 6. Impedance is made up of

(A) Resistance and inductance

(B) Resistance and capacitance

(C) Resistance and inductance or capacitance

(D) Resistance and reactance.

Q. 7. Choose the correct expression

(A) $Z = \sqrt{R^2 + X^2}$ (B) $Z = R^2 + X^2$

(C) $Z = \sqrt{\frac{R^2}{X^2}}$ (D) $\cos\phi = \frac{R}{X}$

Q. 8. To have minimum loss, we would prefer

(A) Aluminium wire (B) Iron wire

(C) Silver wire (D) None of these.

Q. 9. Fixed resistors are mostly made in the form of

(A) Carbon resistors (B) Wire-wound resistors

(C) Thin-film resistors (D) None of these.

Q. 10. Higher inductance means

(A) Higher resistance to the flow of alternating current

(B) Lesser resistance to the flow of direct current

(C) Lesser resistance to the flow of AC current

(D) None of these.

Q. 11. Inductance (L) is given by which of the following ?

(A) $L = \frac{\mu A N^2}{l}$ (B) $L = \frac{\mu N^2}{A l}$

(C) $L = \mu A N^2 l$ (D) None of these.

Q. 12. Voltage across an inductor generally

(A) Leads the current by 90°

(B) Lags the current by 90°

(C) Is inphase with the current

(D) None of these.

Q. 13. In the circuit shown, current I is equal to

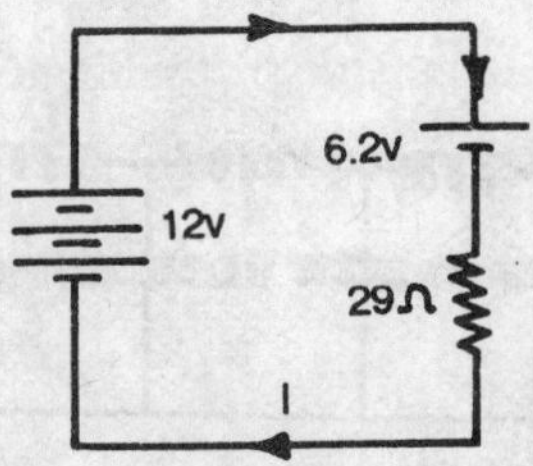

Fig. 1.2

(A) 200 mA (B) ≈ 620 mA

(C) ≈400 mA (D) None of these.

Q. 14. In the circuit shown, current I is

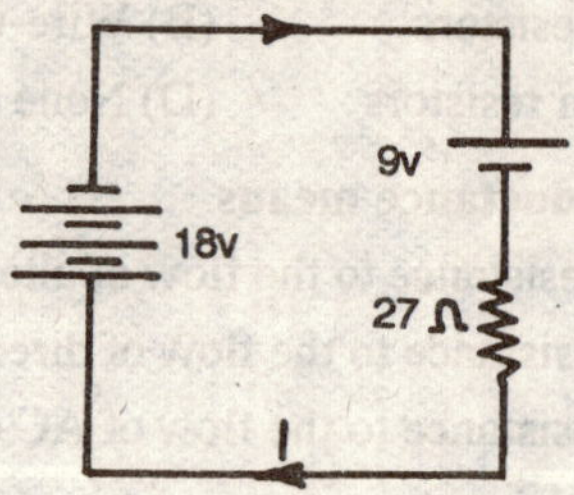

Fig. 1.3

(A) + 1 A (B) - 1 A

(C) - 0.66 A (D) 0.33 A.

Q. 15. A certain source V volts with impedance (A +jB) Ω is driving such a load as to have maximum power transfer. Power transferred is given by

(A) $\frac{V^2}{A}$ watts (B) $\frac{V^2}{2A}$ watts

(C) $\frac{V^2}{4A}$ watts (D) Indeterminate.

Q. 16. In the circuit shown, for maximum power transfer during discharge, lamp impedance should be equal to

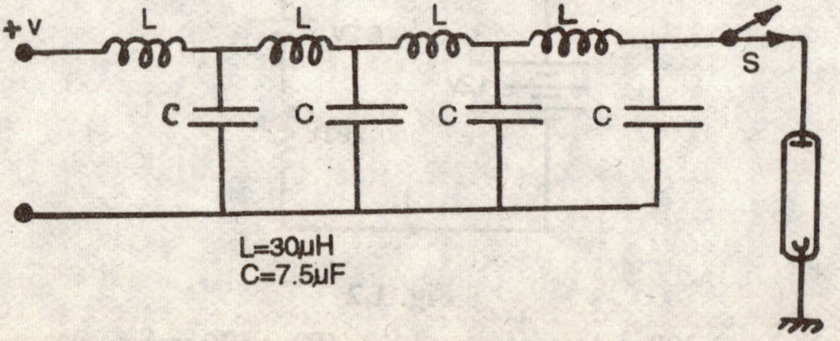

Fig. 1.4

(A) ≈ 4 Ω (B) ≈2 Ω

(C) ≈1 Ω (D) None of these.

Q. 17. In Fig. 1.5 voltage across X-X is equal to

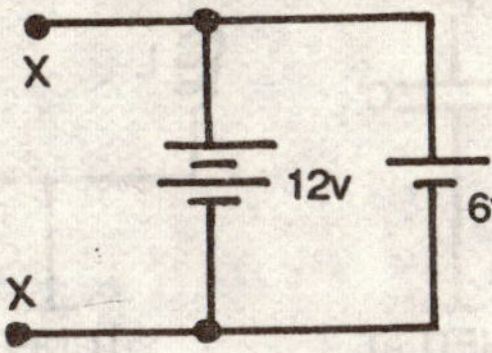

Fig. 1.5

(A) 6 V (B) 12 V

(C) 18 V (D) None of these.

Q. 18. In the network shown the signal at point B contains which of the following frequency components ?

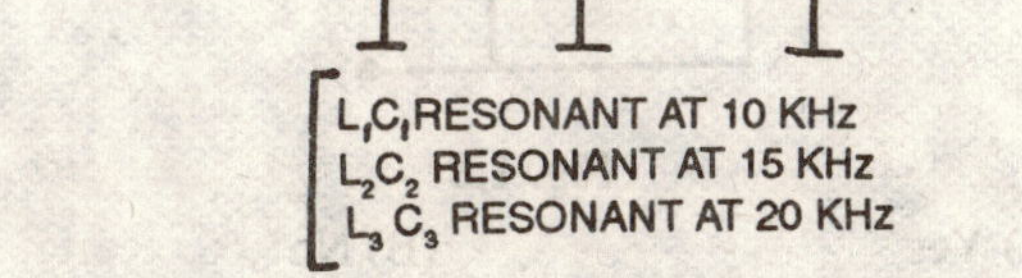

Fig. 1.6

(A) 5 KHz (B) 10, 15 and 20 KHz

(C) No signal (D) None of these.

Q. 19. In the network shown, which has the maximum amplitude at the output point (γ) ?

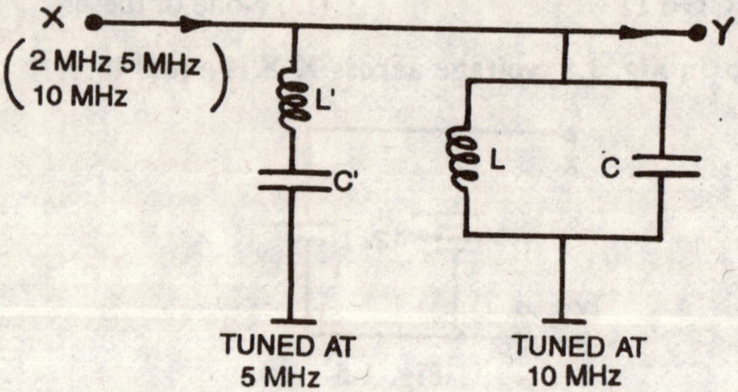

Fig. 1.7

(A) 5 MHz (B) 2 MHz

(C) 10 MHz (D) None of these.

Q. 20. Does network shown represent a possible Thevenin's Equivalent Circuit.

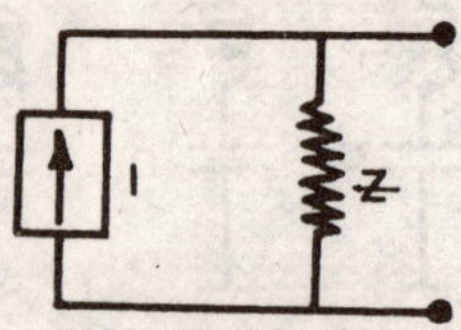

Fig. 1.8

(A) Yes (B) No

(C) Neither of the two.

Q. 21. Norton's Theorem is applicable to circuits with

(A) D.C. sources only

(B) A.C. sources only

(C) D.C. or A.C. or a combination of both

(D) None of these.

Q. 22. At resonance, impedance is

(A) Purely resistive
(B) Purely inductive
(C) Purely capacitive
(D) None of these.

Q. 23. A parallel resonant circuit is formed from $\lambda = 10\mu H$ and $C = 30\ \mu H$ having an associated resistances of $20\ \Omega$ and 10Ω respectively. Impedance at resonance is equal to

(A) $10\ \Omega$
(B) Infinite
(C) $6.66\ \Omega$
(D) None of these.

Q. 24. Increasing circuit's R

(A) Has no effect on bandwidth
(B) Decreases its bandwidth
(C) Increases its band-width
(D) None of these.

Q. 25. Bandwidth in a parallel resonant circuit is given by

(A) $BW = \frac{f_0}{Q_0}$
(B) $BW = \frac{Q_0}{f_0}$
(C) $BW = \frac{R}{4\pi L}$
(D) None of these.

ANSWERS

1. A,	2. A,	3. A,	4. A,	5. B,	6. D,
7. A,	8. C,	9. A,	10. A,	11. A,	12. A,
13. A,	14. B,	15. C,	16. B,	17. B,	18. A,
19. C,	20. B,	21. C,	22. A,	23. D,	24. C,
25. A.					

Chapter 2

Transistor

STUDY MATERIAL

Transistors have replaced bulky vaccum tubes. They offer several advantages over tubes. There are as under:

1. Smaller in size
2. Light in weight
3. No heater or filament needed
4. Usually operate on low voltages
5. Consume little power
6. Improved circuit efficiency
7. Long life
8. Shock proof.

Transistor. It is a silicon or germanium crystal containing three separate regions. Transistors may be NPN or PNP type both having three regions. The middle region is known as base and the two outer regions are known as the emitter and the collector.

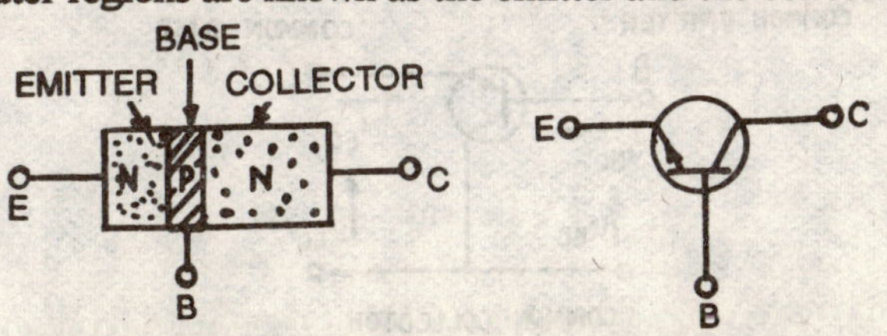

PNP Transistors NPN Transistor

Fig. 2.1 A

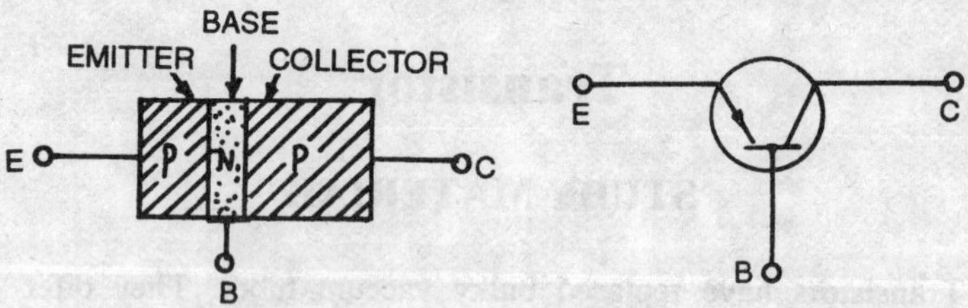

Fig. 2.1 B

Transistor Configurations

In case of transistors any one of the three terminals can be selected as reference or common terminal. Emitter is generally the common terminal. The voltage drops are positive going from the non-reference terminals to the reference terminal. NPN transistor connections are shown in Fig. 2.2.

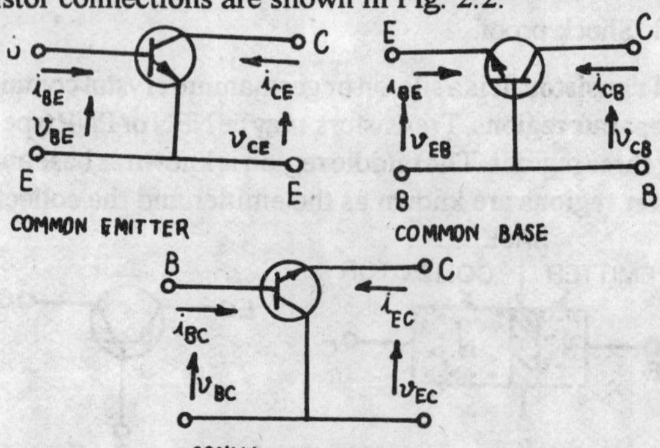

Fig. 2.2. NPN Transistore Connections.

Transistor connection	*Positive voltages*	*Positive currents*	*Signal inputs*
Common emitter CE	V_{BE} and V_{CE}	i_{BE} and i_{CE}	Base
Common base CB	V_{CB}	i_{CB}	Emitter
Common Collector CC	V_{BC}	i_{BC}	Base

In case of PNP transistor all the actual instantaneous voltages and currents have signs, which are opposite to the corresponding ones for NPN transistors.

Common Emitter Amplifier

A common emitter amplifier using PNP transistor is shown in Fig. 2. 3. As the emitter is common, both power supplies have the same polarity. With PNP transistor all the transistor voltages and currents are negative in the common emitter connection.

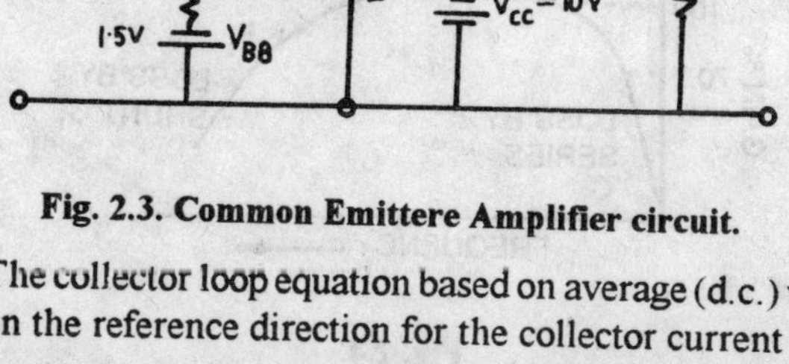

Fig. 2.3. Common Emittere Amplifier circuit.

The collector loop equation based on average (d.c.) values going in the reference direction for the collector current is

$$+ V_{CC} + R_L I_{CE} + V_{CE} = 0$$

or $$-\left(\frac{+R_L}{V_{CC}}\right) V_{CE} + \left(\frac{1}{V_{CC}}\right) V_{CE} = +1$$

which is a straight line

RC Coupled Amplifier

This amplifier is often used for audio frequency amplifiers because R_L permits the same amount of gain for a wide range of audio frequencies. The flat frequency response of the amplifier is also shown in Fig. 2.4. At low frequencies, the output is decreased because of the reactance of C_1. In this X_C increases as the frequency decreases. Larger values for both C_1 and R_1 improve the low frequency response.

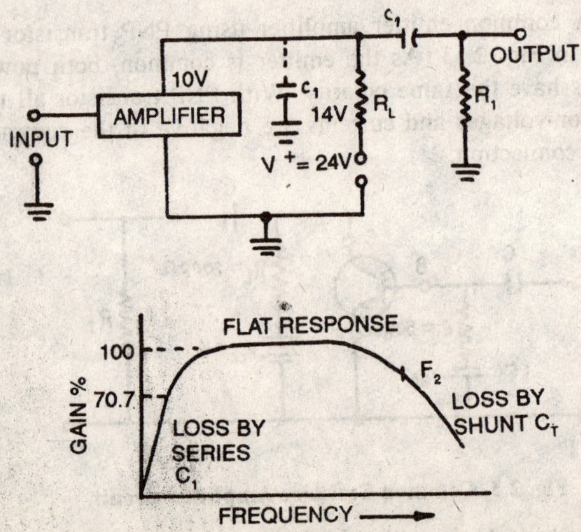

Fig. 2.4.

At high frequencies, the amplifier gain decreases because of the shunting effect of the stray distributed capacitances. The smaller values of C_T improve the high frequency response. Furthermore, smaller R_L values also extend the high frequency gain for flat response, although the overall gain is reduced.

Transformer Coupling

In this the audio output stage is provided with transformer coupling to drive a loud speaker. In Fig. 2.5. L_P provides the primary load impedance for the amplifier. Its d.c. resistance being load, it allows most of the V^+ supply to be used as the amplifier. The a.c. component of the signal current in L_P induces the desired signal voltage in L_S by transformer action. A_S L_S is an isolated winding the d.c. component of the primary voltage and current is blocked. No coupling capacitor is required in this case.

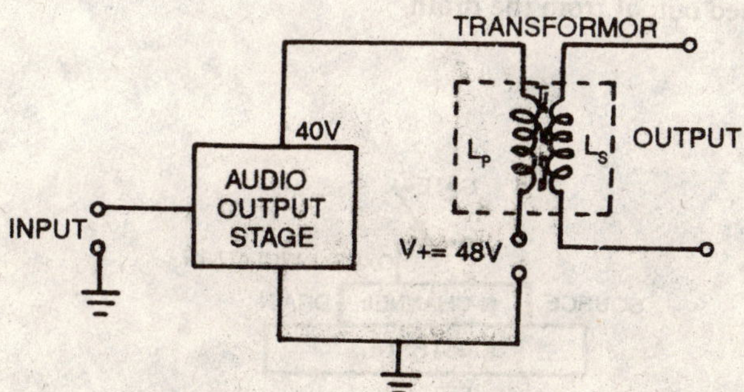

Fig. 2.5.

A transformer that couples signals between two amplifiers is known as an interstage transformer.

Field Effect Transistor (FET)

This semiconductor device is similar to a transistor, provided with high input resistance (in megaohms). It can take several volts for input circuit, compared to fraction of a volt in case of transistors. The FET operation depends on controlling current through a semi-conductor channel of one polarity. In this bulk of the substrate material is neutral or lightly doped silicon. The substrate material serves as a substrate on which the other electrodes are diffused.

FET is a unipolar device with high resistance from gate to source. In this the charge carrier enter the channel bar to provide current through the channel. The current leaves the channel through the drain. Gate controls the conductance of the channel between the source and the drain. Thus gate corresponds to the base, drain to the collector and source to the emitter.

The input signal is generally applied to the gate, with amplified output from the drain.

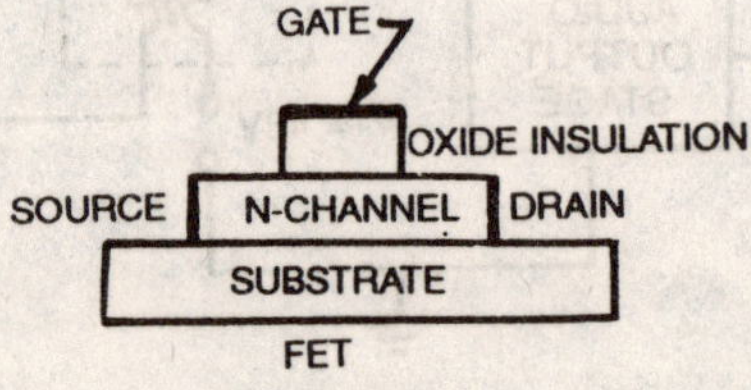

Fig. 2.6. A Symbol for JFET.

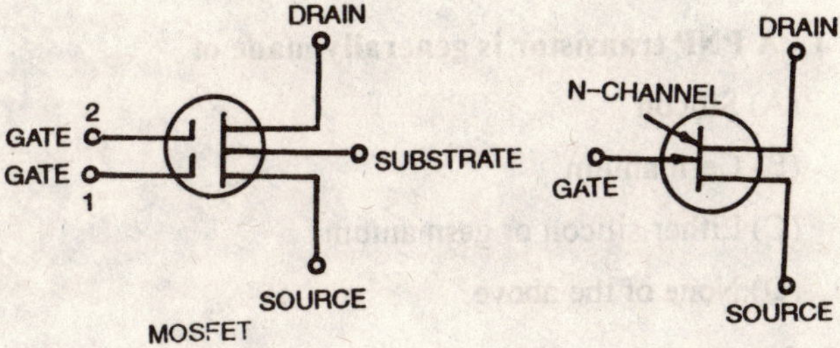

Fig. 2.6 B. Symbol for MOSFET with double gate.

JFET

In junction field effect transistor, the gate and channel form a conventional PN junction. However, this junction has high input resistance because of reverse bias.

IGFET

In this the insulted gate consists of a metal electrode separated from the channel by a thin layer of glassy silicon dioxide. Voltage applied to the gate can induce charges in the channel to control the drain current. The insulated gate has very high input resistance independent of the polarity of the gate voltage. This is a metal oxide semiconductor field effect transistor MOSFET or an IGFET. This is widely used in low power amplifiers.

OBJECTIVE TYPE QUESTIONS

Q. 1. A PNP transistor is generally made of

(A) Silicon

(B) Germanium

(C) Either silicon or germanium

(D) None of the above.

Q. 2. In a transistor the region that is very lightly doped and is very thin is

(A) Emitter (B) Base

(C) Collector (D) None of the above.

Q. 3. In a NPN transistor, when emitter junction is forward biased and collector junction is reverse biased, the transistor will operate in

(A) Active region (B) Saturation region

(C) Cut of region (D) Inverted region.

Q. 4. A transistor will operate in inverted region if

(A) Emitter junction is forward biased and collector junction is reverse biased

(B) Emitter junction is reverse biased and collector junction is forward biased

(C) Emitter junction as well as collector junction are forward biased

(D) Emitter junction as well as collector junction are reverse biased.

Q. 5. Choose the correct statement:

(A) FET and junction transistor both are unipolar

(B) FET and junction transistor both are bipolar

(C) The FET is bipolar, while junction transistors are unipolar

(D) The FET is unipolar, while junction transistors are bipolar.

Q. 6. The encapsulation of transistor is essential for

(A) Preventing radio interference

(B) Preventing photo-emission effects

(C) Avoiding loss of free electrons

(D) Mechanical ruggedness.

Q. 7. A diac is a semiconductor device that acts as a

(A) 2 terminal unidirectional switch

(B) 2 terminal bidirectional switch

(C) 3 terminal bidirectional switch

(D) 4 terminal multidirectional switch.

Q. 8. Which of the following is essentially for transistor action ?

(A) The base region must be very wide

(B) The base region must be very narrow

(C) The base region must be made of some insulating material

(D) The collector region must be heavily doped.

Q. 9. In a transistor, current I_{CBO} flows when

(A) Some d.c. voltage is applied in the reverse direction to the emitter junction with the collector open circuited

(B) Some d.c. voltage is applied in the forward direction to the collector junction with the emitter open circuited

(C) Some d.c. voltage is applied in the reverse direction to the collector junction with the emitter open circuited

(D) Some d.c. voltage is applied in the forward direction to the emitter junction with the collector open-circuited.

Q. 10. The current I_{CBO}

(A) Increases with increase in temperature

(B) Is normally greater for silicon transistors than germanium transistors

(C) Mainly depends on the emitter base junction bias

(D) Depends largely on the emitter doping.

Q. 11. In a transistor α is related to β by the relation

(A) $\beta = \frac{\alpha + 1}{\alpha}$ (B) $\beta = \frac{\alpha - 1}{\alpha}$

(C) $\beta = - \frac{\alpha}{\alpha - 1}$ (D) $\beta = \frac{\alpha}{\alpha + 1}$

Q. 12. Thermal runaway of a transistor occurs when

(A) Heat dissipation from transistor is excessive

(B) Transistor joints melt due to high temperature

(C) There is excessive leakage current due temperature rise

(D) None of the above.

Q. 13. As compared to a CB amplifier, a CE amplifier has

(A) Lower current amplification

(B) Higher current amplification

(C) Lower input resistance

(D) Higher input resistance.

Q. 14. In Fig. 2.7 the source bias voltage is equal to $I_s R_s$ for an FET. What is the gate-source bias V_{GS} ?

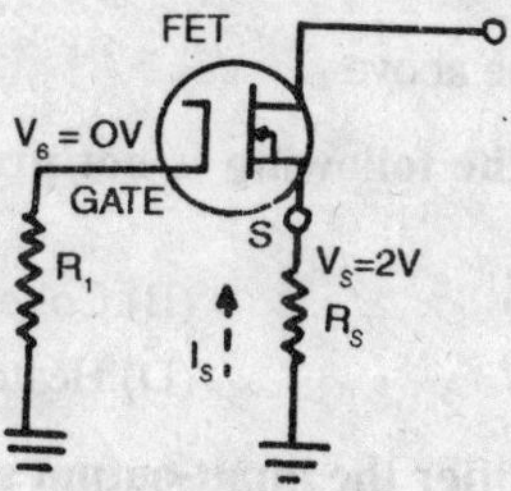

Fig. 2.7.

(A) 1.04 V (B) 1.66 V

(C) 2.04 V (D) 2.66 V.

Q. 15. Which of the following circuit does not require a coupling capacitor ?

(A) Resistance loaded (B) Transformer coupled

(C) Impedance coupled (D) Single tuned.

Q. 16. In an RC-coupled amplifier for improving the low frequency response

(A) Lower R_L is provided (B) More bias is used

(C) Less gain is provided (D) Higher C_C is used.

Q. 17. Which of the following statement is correct ?

(A) The emitter injects holes into the base region of the PNP transistor and electrons into the base region of the NPN transistor

(B) The emitter injects electrons into the base region of the PNP transistor and holes into the base region of the NPN transistor

(C) The emitter junction is reverse biased in the NPN transistors and forward biased in NPN transistor

(D) None of the above.

Q. 18. Which of the following is not provided in a PNP transistor ?

(A) Base (B) Collector

(C) Emitter (D) Heater.

Q. 19. For an amplifier the input-output signals are shown in Fig. 2.8. The class of amplifier is

(A) A (B) AB

(C) B (D) C.

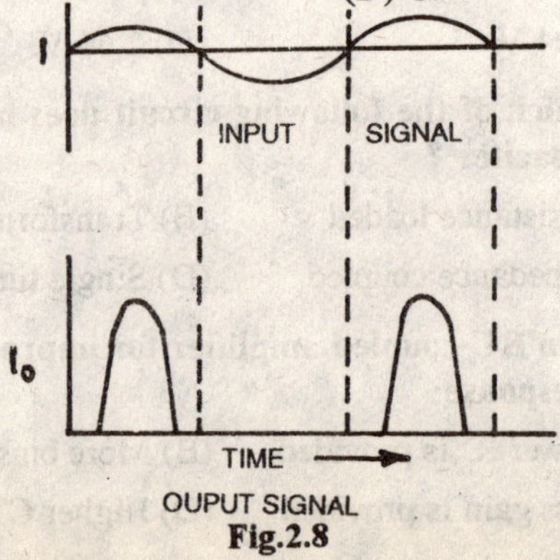

Fig.2.8

Q. 20. Which coupling produces the minimum interference with frequency response ?

(A) RC coupling (B) Transformer coupling

(C) Direct coupling (D) Impedance coupling.

Q. 21. For an amplifier the power gain in decibels will be equal to voltage gain in decibels only when

(A) $R_o = 0$ (B) $R_o = R_i = 0$

(C) $R_o = R_i$ (D) $R_i = 0$.

Q. 22. Which of the following is the fastest switching device?

(A) JFET (B) BJT

(C) MOSFET (D) Triode.

Q. 23. Which of the following statement is false ?

(A) The zener diode is used as a constant voltage source

(B) The SCR is a silicon rectifier with a gate electrode to control when current flows from cathode to anode

(C) The gate electrode on the FET corresponds to the collector in a bipolar transistor

(D) None of the above.

Q. 24. An n-channel FET is used in the amplifier shown in Fig. 2.9. The transistor parameters are I_{DSS} = 5 mA and V_P = — 4V. The amplifier is to be designed so that V_{DS} = 8 volts when $V_1 = V_{GS} = 0$ and the power-supply voltage is to be V_{DD} = 25 volts.

What should be the value of Rd ?

(A) 3.4Ω (B) 34Ω

(C) 340Ω (D) 3400Ω.

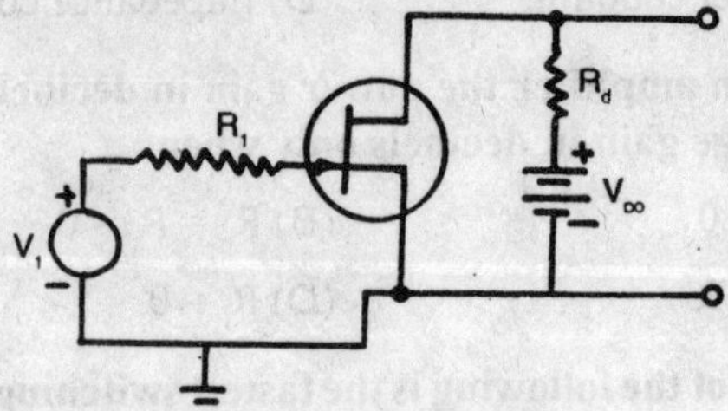

Fig. 2.9.

Q. 25. A BJT used in the circuit of Fig. 2.10 A produces the base-current waveform shown in Fig. 2.10 B when the switch is thrown from position 1 to position 2.

Assuming that negligible charge is lost by recombination during the 0.1 μ sec turn off interval, estimate the amount of charge that was stored in the transistor just before the switch was thrown

(A) 1 pcoul (B) 5 pcoul

(C) 10 pcoul (D) 100 pcoul.

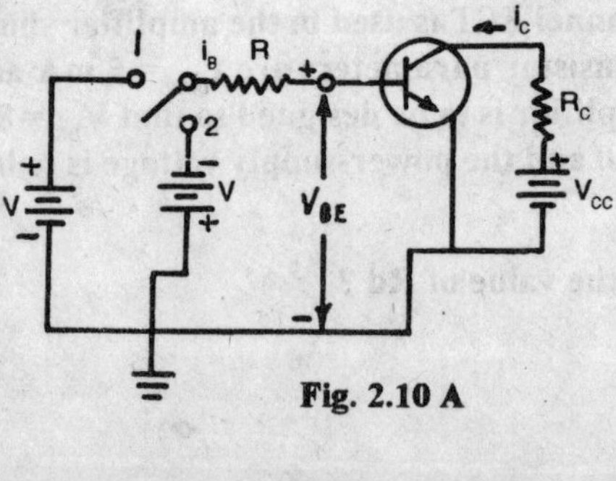

Fig. 2.10 A

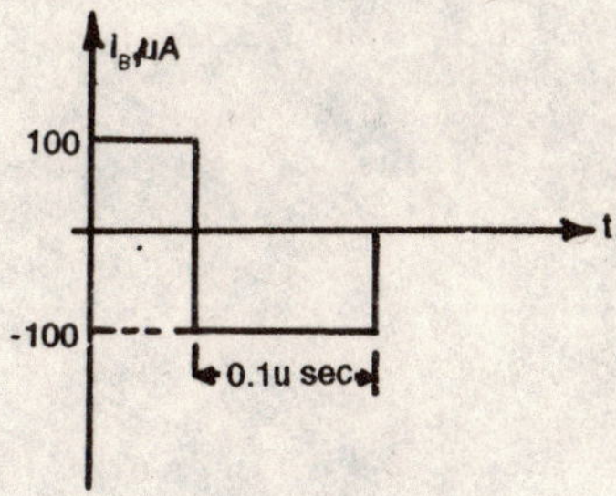

Fig. 1.10 B

ANSWERS

1. C,	2. B,	3. A,	4. B,	5. D,	6. D,
7. A,	8. B,	9. C,	10. A,	11. C,	12. C,
13. B,	14. A,	15. B,	16. D,	17. A,	18. D,
19. D,	20. C,	21. C,	22. C,	23. D,	24. D,
25. C.					

Chapter 3

Modulation

STUDY MATERIAL

Modulation. It is the process wherein some characteristics of a high frequency sine wave (the carrier) is changed in accordance with the instantaneous value of the modulating signal (the audio). The instantaneous value of the sine wave carrier is given by equation $e_c = E_c \sin (W_c t + \theta)$, in this E_c is the maximum amplitude, W_c is the angular velocity or frequency, while θ is the phase angle with respect to some reference. Any of these three quantity, i.e., amplitude, frequency and phase are changed in accordance with modulating signal.. Various types of modulation methods are as under.

(i) Amplitude modulation, (ii) Frequency modulation, (ii) Phase modulation, (iv) Pulse modulation.

It is impossible to transmit modulating signal or audio signal, because for efficient radiation and reception the transmitting antenna must have heights at-least $\lambda/4$ wavelength. If we want to transmit a frequency of 15KHz then vertical antenna of height 5000 Ms will be required which is practical impossible. Then carrier is needed, which is high frequency. The modulating signal will be super-imposed upon the carrier. Let carrier frequency be 1 MHz then antenna height of $\lambda/4$ wavelength will be 75 Ms, which can be easily erected.

Also it is essential to translate audio frequency into different portions of electromagnetic waves for that carrier frequency is required. This also overcomes the difficulties of poor radiation at low frequency.

Expression for Amplitude Modulation

In a system of modulation wherein the amplitude of the carrier is made proportional to instantaneous amplitude of the modulating voltage

Let the carrier voltage be given by

$$e_c = E_c \sin w_c t$$

and the modulating voltage be given by

$$e_m = E_m \sin w_m t$$

Then modulation index, $m = \dfrac{E_m}{E_c}$

The value of m lies between 0 and 1

The equation for the amplitude of the modulated voltage will be given by

$$A = E_c + e_m = E_m + E_m \sin w_m t = E_c (1 + m \sin w_m t)$$

The voltage instantaneous of the resulting amplitude-modulated wave is given by

$$e = E_c \sin w_c t + \frac{m E_c}{2} \cos (w_c — w_m) t$$

$$- \frac{m E_c}{2} \cos (w_c + w_m) t.$$

Frequency Spectrum of AM

AM is shown simply as made up of three discrete frequencies. Of these, the central frequency i.e., carrier has the highest amplitude. Two side band frequencies. (Lower sideband LSB and Upper sideband USB) are disposed symmetrically about it, having amplitudes which are equal to each other, but which can never exceed half the carrier amplitude.

$$\text{Thus}\quad E_m = m\,E_c = \frac{E_{max} - E_{min}}{2}$$

$$\text{Hence } m = \frac{E_{max} - E_{min}}{E_{max} + E_{min}}$$

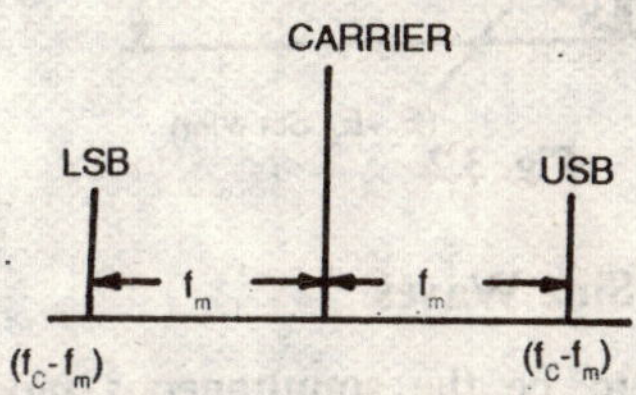

Fig. 3.1.

Power Relation in AM Waves

The total power of modulate wave is given by

$$P_t = \frac{(E_{carr})^2}{R} + \frac{(E_{LSB})^2}{R} + \frac{(E_{USB})^2}{R}.$$

Where all three voltages are r.m.s. values and R is the resistance of antenna, wherein the power is dissipated.

The unmodulated carrier power is given by

$$P_c = \frac{(E_{carr})^2}{R} = \frac{(E_c/\sqrt{2})^2}{R} = \frac{E_c^{\,2}}{2R}$$

$$\text{Also } P_{LBS} = P_{USB} = \frac{m^2}{2}\;\frac{E_c^{\,2}}{2R}$$

$$\text{Hence,}\quad \frac{P_t}{P_c} = 1 + \frac{m^2}{2}$$

and for currents $\frac{I_t}{I_c} = \sqrt{1 + \frac{m^2}{2}}$

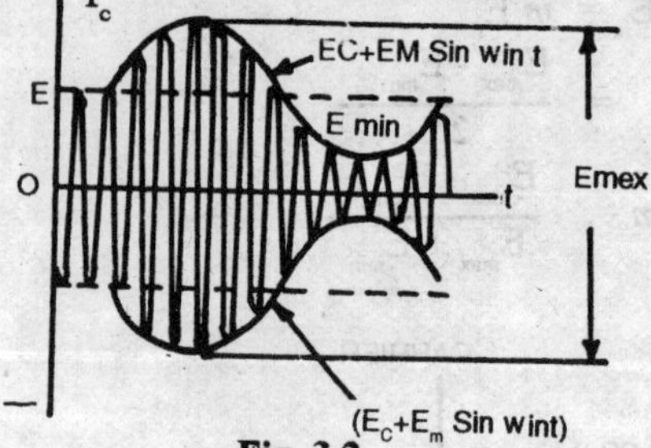

Fig. 3.2.

Modulation by Several Sine Waves

Let E_1, E_2, E_3, etc. be the simultaneous modulating voltages. Then total modulating voltage is given by

$$E_t = \sqrt{E_1^2 + E_2^2 + E_3^2 + ...}$$

and $$\frac{E}{E_c} = \sqrt{\frac{E_1^2}{E_c^2} + \frac{E_2^2}{E_c^2} + \frac{E_3^2}{E_c^2} + ...}$$

so that $m_1 = \sqrt{m_1^2 + m_2^2 + m_3^2 + ...}$

Frequency Modulation

An unmodulated wave may be written as

$$x = A \sin (wt + \phi)$$

In this when w is changed the modulation is called frequency modulation, the amount whereby the carrier frequency is varied from the unmodulated value is known as the deviation. The rate at which this frequency variation or oscillation occurs is equal to the modulating frequency. The instantaneous frequency f of the frequency modulated wave is given by

$$f = f_c(1 + k\,E_m \cos w_m t)$$

Where k = proportionality constant

$E_m \cos w_m t$ = instantaneous modulating voltage

The instantaneous frequency is given by $f = f_c(1 \pm k\,E_m)$ and maximum deviation $\delta = k\,E_m f_c$.

The modulation index for F.M., m_f is given by

$$m_f = \frac{\text{Maximum frequency deviation}}{\text{Modulating frequency}} = \frac{\delta}{f_m}$$

Some points that may be noted for FM are:

(1) In AM there are only three frequencies—the carrier, LSB and USB. In FM, there is an infinite number of side *b* and as well as the carrier. They are separated from the carrier by f_m, $2f_m$, $3f_m$,... with a recurrence frequency of f_m.

(2) The sidebands at equal distances from f_c have equal amplitudes, so that the side band distribution is symmetrical about the carrier frequency.

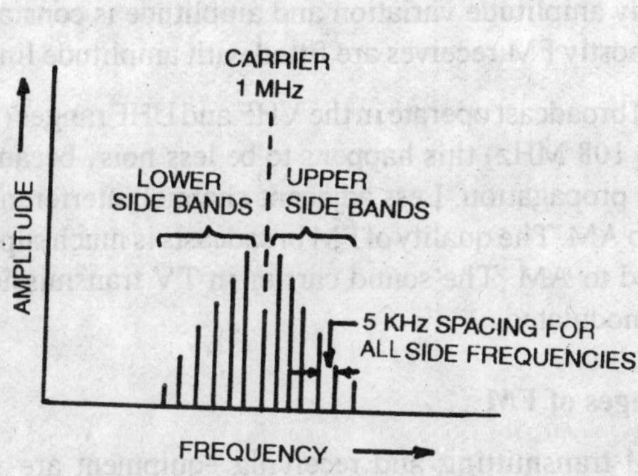

Fig. 3.3.

(3) In AM increased depth for modulation increases the sideband power, and therefore the total transmitted power. In FM, the total transmitted power always remains constant, but with increased depth of modulation the required bandwidth is increased.

(4) The theoretical bandwidth needed in FM is infinite. In practice, the bandwidth used is one that has been calculated to allow for all significant amplitudes of sideband components under the most exacting conditions.

(5) In FM, unlike in AM, the amplitude of the carrier component does not remain constant.

(6) It is possible for the carrier component of the frequency modulated wave to disappear completely. This happens for certain so-called eigen values of the modulation index.

Advantages of FM over AM

1. FM reception is less noisy in comparison to AM.

This is because the man made noises atmospheric noises produce only amplitude variation and amplitude is constant in FM. Also mostly FM receives are fitted with amplitude limiter.

2. FM broadcast operate in the VHF and UHF ranges (VHF range 88 to 108 MHz) this happens to be less noisy because of space-wave propagation. Less adjacent channel interference as compared to AM. The quality of FM broadcasts is much superior as compared to AM. The sound carrier in TV transmission is frequency modulated.

Disadvantages of FM

1. FM transmitting and receiving equipment are more complicated. Much costly as compared to AM.

2. The area of reception is limited only upto line of sight.

3. A much wider channel is required for FM.

Phase Modulation

In the equation for unmodulated wave, $e = A \sin(w_c t + \phi)$, where ϕ is changed the system of modulation is known as the *phase modulation*. The expression for phase modulated wave is given as, $e = A \sin(w_c t + \phi_m \sin w_m t)$

Where ϕ is the maximum value of phase change introduced by this particular modulating signal and is proportional to the maximum amplitude of this modulation.

In phase modulation, the phase deviation is proportional to the amplitude of the modulating signal and therefore independent of its frequency.

Pulse Modulation

A pulse transmitter is turned on and off at a regular rate to transmit pulses of RF energy. The ratio of 'time on' to 'time off' is called the duty cycle. Without any modulation, all the transmitted pulses have the same amplitude, width and spacing. However, the modulating intelligence can be used to control one characteristic of the pulses. Various methods of pulse modulation are:

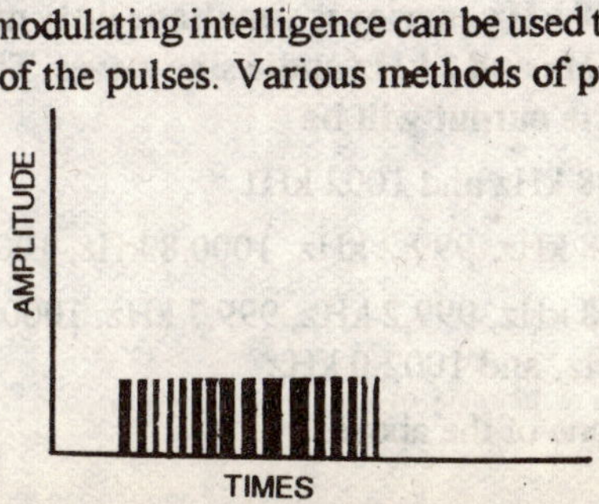

Fig. 3.4. Pulse width Modulation (PWM)

(1) Pulse Amplitude Modulation, PAM

(2) Pulse Width Modulation, PWM

(3) Pulse Frequency Modulation, PFM

(4) Pulse Coded Modulation, PCM.

Pulse modulation allows high transmitter efficiency and good signal to noise ratio, but it requires appreciable bandwidth.

OBJECTIVE TYPE QUESTIONS

Q. 1. Fig. 3.5 below an example of

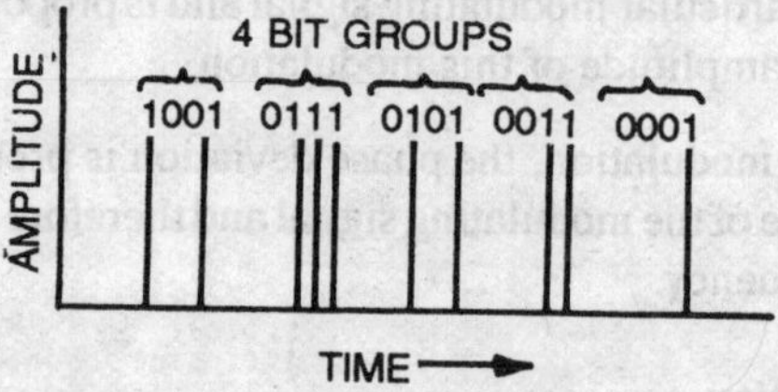

Fig. 3.5.

(A) PAM (B) PWM

(C) PFM (D) PCM.

Q. 2. A 1000 kHz carrier is simultaneously modulated with 300 Hz, 800 Hz and 2 kHz audio sine waves. The frequencies present in the output will be

(A) 998 kHz and 1002 kH1

(B) 998 kHz, 999.2 kHz, 1000.8 kHz, 1002.0 kHz

(C) 998 kHz, 999.2 kHz, 999.7 kHz, 1000.3 kHz, 1000.8 kHz, and 1002.0 kHz

(D) None of the above.

Q. 3. The percentage of modulation for 45 kHz deviation in the FM broadcast band will be

(A) 30% (B) 40W

(C) 60% (D) 90%.

Q. 4. For a FM wave carrier modulating frequency is 10 kHz and bandwidth is 2 MHz. If the modulating signal amplitude is doubled, the bandwidth will be

(A) 0.5 MHz (B) 1 MHz

(C) 2 MHz (D) 4 MHz.

Q. 5. In AM receiver the oscillator frequency is

(A) Always equal to signal frequency

(B) Always equal to 455 kHz

(C) Always higher than signal frequency

(D) Always lower than signal frequency.

Q. 6. Select the one in which quantization noise occurs

(A) PCM (B) TDM

(C) FDM (D) PWD.

Q. 7. Select the statement that is false.

(A) A stationary process is always a ergodic process

(B) A ergodic process is always a stationary random process

(C) It is possible to have a stationary random process that is not ergodic

(D) None of the above.

Q. 8. If Pk be the probability of the message to be received or transmitted, then the amount of information (I_k) associated, in bits, will be given by

(A) $I_k = \log_2 P_k$ (B) $I_k = \log_2 \frac{1}{P_k}$

(C) $I_k = \frac{1}{\log_2 P_k}$ (D) $I_k = \frac{1}{\log_2 \frac{1}{P_k}}$.

Q. 9. In grid bias modulation system the power needed for modulating amplifier.

(A) is negligibly low since grid is negative

(B) is large as compared to that for plate modulation

(C) is almost the same as compared with plate modulation

(D) is small as compared to that for plate modulation.

Q. 10. The typical squelch circuit

(A) Cuts off an audio amplifier when the carrier is absent

(B) Cuts off an IF amplifier when the AGC is maximum

(C) Cuts off an IF amplifier when the AGC is minimum

(D) Eliminates the RF interference when the signal is weak.

Q. 11. For a signal amplitude modulated to a depth of 100% by a sinusoidal signal, what is the power ?

(A) Same as the power of unmodulated carrier

(B) Twice as the power of unmodulated carrier

(C) Four times the power of unmodulated carrier

(D) 3/2 times the power of unmodulated carrier.

Q. 12. In FM, the frequency deviation is generally

(A) Proportional to modulating frequency

(B) Proportional to amplitude of modulating signal

(C) Constant

(D) Directly proportional to amplitude and inversely proportional to modulating frequency.

Q. 13. Audio frequency range lies between

(A) 20 Hz and 20 kHz (B) 20 kHz and 200 kHz

(C) 2 MHz and 20 MHz (D) 20 MHz and 200 MHz.

Q. 14. Maximum undistorted power output of a transmitter can be expected when its modulation is

(A) 50% (B) Between 50% and 90%

(C) 100% (D) More than 100%.

Q. 15. An FM signal with a deviation α is passed through a mixer, and has its frequency reduced six fold. The deviation in the output of the mixer will be equal to

(A) 6α (B) $\sqrt{6\alpha}$

(C) α (D) Indeterminate.

Q. 16. A matching network for coupling the transmitter output to the antenna is shown in Fig. 3.6.

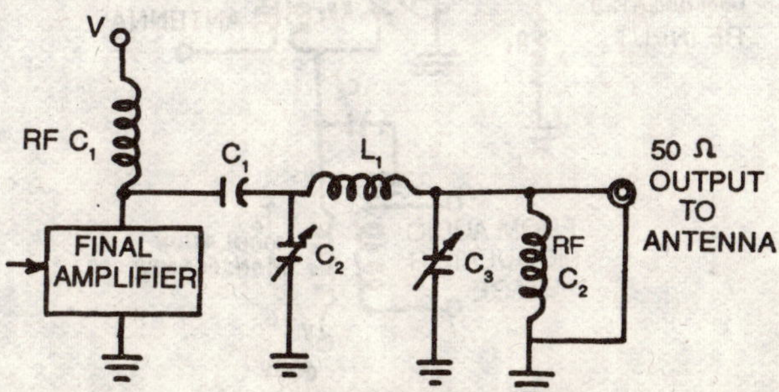

Fig. 3.6.

In this C_2 should be tuned for

(A) Minimum d.c. plate current

(B) Maximum d.c. plate current

(C) Maximum RF antenna current

(D) Specified value of d.c. plate current.

Q. 17. In the stabilized reactance modulator AFC system

(A) The discriminator must have a fast time constant to prevent demodulation

(B) The higher the discriminator frequency, the better the oscillator frequency stability

(C) The discriminator frequency must not be too low, or the system will fail

(D) Phase modulation is converted into FM by the equalizer circuit.

Q. 18. What is the RF amplifier plate voltage when the audio-modulating voltage is + 600 V ?

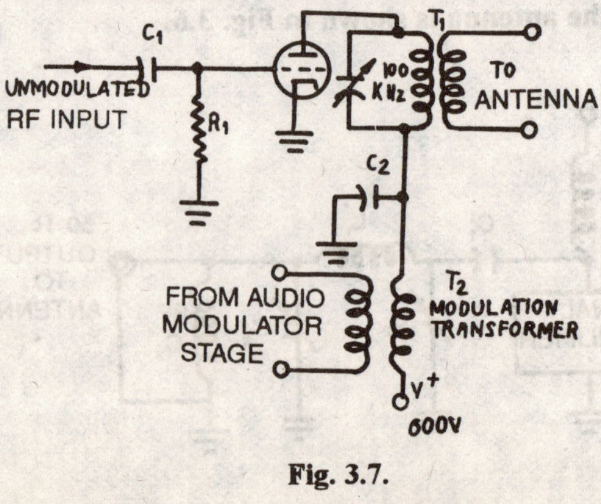

Fig. 3.7.

(A) - 600 V (B) Zero

(C) 600 V (D) 1200 V.

Q. 19. The maximum I or V on the modulated carrier wave is 5 units and the minimum is 3 units. The percentage of modulation will be equal to

(A) 100 (B) 80

(C) 50 (D) 25.

Q. 20. Which of the following statement is NOT correct ?

(A) In frequency modulation the loudest sounds produce maximum frequency deviation

(B) Modulation is used to reduce bandwidth

(C) In FM, the audio modulating frequency determines the rate of frequency swing

(D) None of the above.

Q. 21. An oscillator at 4.2 MHz is followed by two frequency doublers and two triplers. The output frequency will be

(A) 84 MHz (B) 112.4 MHz

(C) 151.2 MHz (D) 303.4 MHz.

Q. 22. The signals may be

I. AM signals

II. Phase modulated signals

III. Frequency modulated signals

Armstrong modulator generates

(A) I and II only (B) II and III only

(C) I and III only (D) I, II and III.

Q. 23. A 3A modulation is sometimes used to

(A) Permit the receiver to have a frequency synthesizer

(B) Reduce the bandwidth required for transmission

(C) Reduce the power that must be transmitted

(D) Simplify the frequency stability problem in reception.

Q. 24. The final power amplifier in an FM transmitter generally operates as

(A) Class A (B) Class B

(C) Class C (D) Class D.

Q. 25. Quantizing noise takes place in

(A) Time-division-multiplex

(B) PCM

(C) PPM

(D) Frequency-division-multiplex.

Q. 26. A pre-emphasis circuit provides extra noise immunity by

(A) Converting the phase modulation to frequency modulation

(B) Preamplifying the whole audio band

(C) Boosting the bass frequencies

(D) Amplifying the higher audio frequencies.

Q. 27. The percentage saving in power of 100% modulated suppressed carrier AM signal is

(A) 80 (B) $66\frac{2}{3}$

(C) 50 (D) 40.

Q. 28. What is the unit of modulation index ?

(A) Hertz (B) $(Hz)^{-1}$

(C) $(Hz)^{-2}$ (D) No unit.

Q. 29. Which of the following statement regarding Armstrong modulator is NOT correct ?

(A) The system is basically phase modulation system

(B) Frequency multiplication is necessarily used

(C) Equalization is unnecessary

(D) A FC is not needed, as a crystal oscillator is used.

ANSWERS

1. D,	2. C,	3. C,	4. D,	5. C,	6. A,
7. A,	8. B,	9. D,	10. A,	11. D,	12. B,
13. A,	14. C,	15. C,	16. A,	17. C,	18. D,
19. D,	20. B,	21. C,	22. B,	23. D,	24. C,
25. B,	26. D,	27. B,	28. D,	29. C.	

Chapter 4

Semi-Conductor Devices

STUDY MATERIAL

A semi-conductor is a material that has a resistance in between that of a conductor and an insulator. A semi-conductor in its purest form is called *intrinsic semi-conductor*. The main features of intrinsic semi-conductors are as under :

(1) R more than metal conductors and less than insulators.

(2) Negative temperature coefficient as R decrease with rise in temperature.

(3) Electron valence of ± 4. This valence means the atom has four electrons in the outermost ring-halfway to the goal of 8.

Doping

The conductivity of semi-conductors can be increased by adding impurity elements. Such a process called *doping*. When the number of free electrons is increased, the doped semi-conductor is negative or N-type; reducing the number of free electrons makes the material P type. Both germanium (Ge) and silicon (Si) can be used with either P-type or N-type doping, but silicon semi-conductor devices are most common.

Silicon and germanium have 5.0×10^{28} and 4.42×10^{28} atoms per m^3 respectively and an addition of only one impurity atom per 10^7 or 10^8 semi-conductor atoms is sufficient to yield the desired properties.

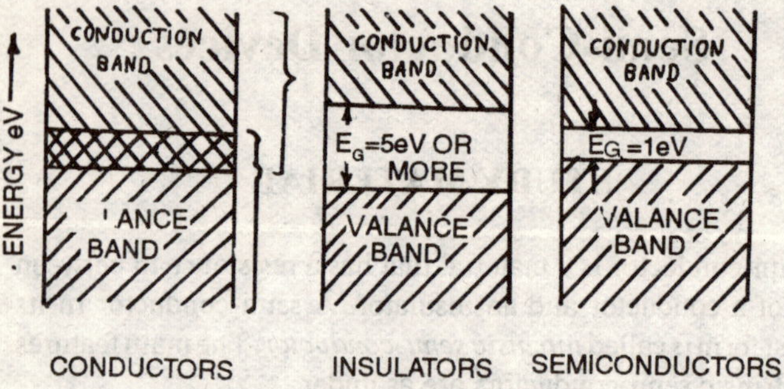

Fig. 4.1.

Semi-Conductors and Doping Elements

Element	*Symbol*	*Atomic Number*	*No. of Valence electrons*
Aluminium	Al	13	3
Boron	B	5	3
Gallium	Ga	31	3
Indium	In	49	3
Germanium	Ge	32	4
Silicon	Si	14	4
Antimony	Sb	51	5
Arsenic	As	33	5
Phosphorous	P	15	5

The doping elements sarenic, antimony and phosphorous have a valence of 5. For each of these atoms there are 5 electrons in the outermost ring. In a covalent bond with Ge or Si atoms having 4 valence electrons each impurity atom providing an extra electron, as shown in Fig. 4.2.

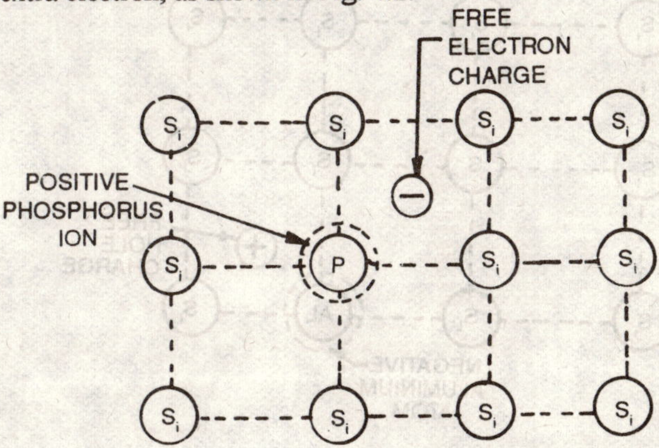

Fig. 4.2.

The doping elements aluminium, boron, gallium, and indium have a valence of 3. For each of these atoms there are 3 electrons in the outermost ring. In a covalent bond with Ge or Si atoms, there are 7 electrons instead of 8 for each bond with an impurity element. The one missing electron in such a covalent bond can be considered as a free positive charge called a hole charge.

Hole Current

A hole has the same amount of positive charge as a proton, equal to an electron but with opposite polarity. However, a hole charge is not a proton. The proton is a stable charge in the nucieus that is not free to move. A hole is a positive charge outside the nucleus present only in semi-conductors because of unfilled covalent bonds. It should be noted, though, that hole current flows only in P-type semi-conductors. The current is electron flow in N-type semi-conductors and all wire conductors.

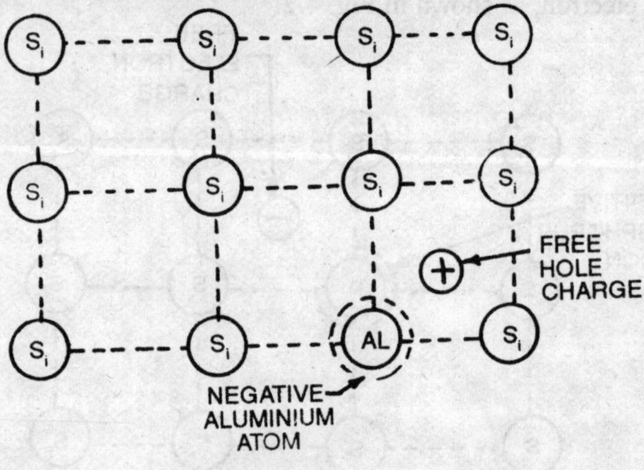

Fig. 4.3.

Current in Semi-conductors

In N-type semi-conductors, free electrons are dominant and are called majority charges. It has minority charges of holes. A P-type semi-conductor has majority hole charges, but there are minority electron charges. When the majority charges are made to move in the semi-conductor by an applied voltage, this result in a relatively large forward current. This is a flow of electrons as the majority carriers in N-type semi-conductors, or hole current in P-type semi-conductors. When minority charges move, this current is in the reverse direction, compared with the forward current of the majority charges. This very small current of minority charges is known as reverse current or leakage current.

Drift Current

Whenever, charges are forced to move by the electric field of a potential difference, drift current is said to flow. For example, the electron flow produced by applied voltage is a drift current.

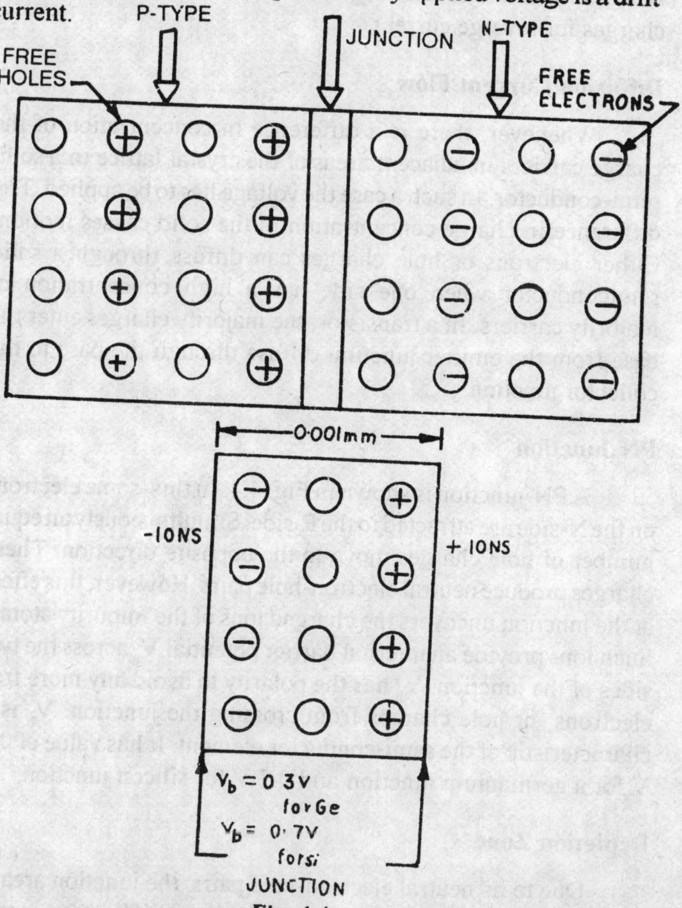

Fig. 4.4.

Hole charges in P-type semi-conductor can also provide a drift current with an applied voltage. However, hole current is in the opposite direction from electron flow. The drift current in semi-conductors can be electron flow or hole current. It can also be a motion of majority charges for forward current or minority charges for reverge current.

Diffusion Current Flow

Whenever, there is a difference in concentration of the charge carriers in adjacent areas of the crystal lattice in a solid semi-conductor. In such a case the voltage has to be applied. The difference in charge concentration in the solid causes motion. Either electrons or hole charges can diffuse through a solid semiconductor when one side has a high concentration of majority carriers. In a transistor, the majority charges enter the base from the emitter junction diffuse through the base to the collector junction.

PN Junction

A PN-junction is shown in Fig. 4.4. In this, some electrons on the N-side are attracted to the P-side. Simultaneously an equal number of hole charges move to the opposite direction. These charges produce neutral electron-hole pairs. However, this effect at the junction uncovers the charged ions of the impurity atoms. Such ions provide an internal barrier potential V_b across the two sides of the junction. V_b has the polarity to avoid any more free electrons or hole charges from crossing the junction. V_b is a characteristic of the semi-conductor element. It has value of 0.3 V for a germanium junction and 0.7 V for silicon junction.

Depletion Zone

Due to its neutral electron-hole pairs, the junction area is the depletion zone, meaning has no free charges. However, it still has the ion charges anchored in position to produce V_b

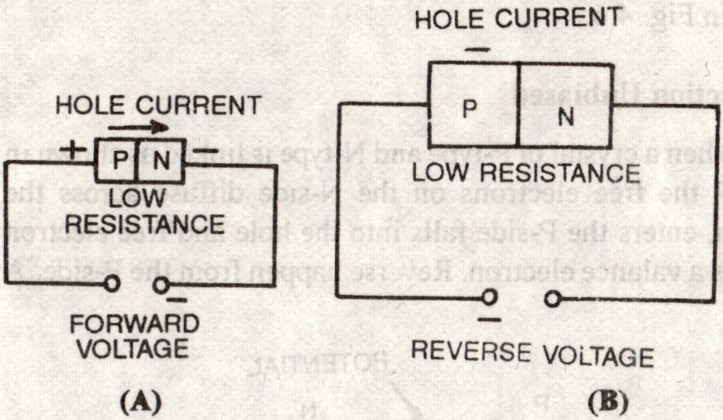

Fig. 4.5.

Forward Voltage

When an external voltage is applied to a PN-junction as shown in Fig. 4.5. A the forward current flows (after neutralising V_b).

Reverse Voltage

When voltage applied to PN-junction is as shown in Fig. 4.5 B a reverse voltage is said to have been applied.

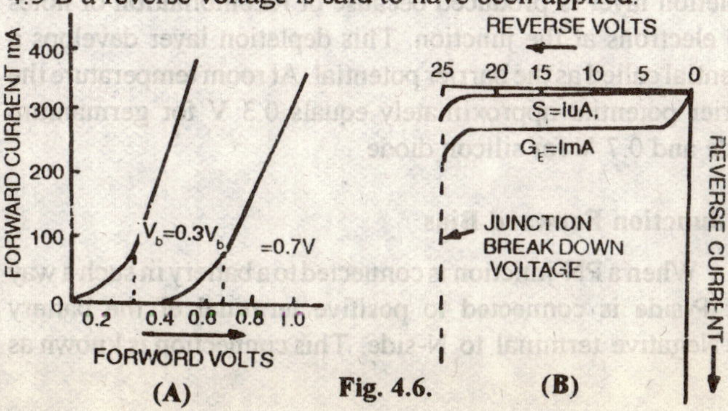

Fig. 4.6.

The VI (volt-ampere) characteristic for a PN-junction is shown in Fig. 4.6.

PN Junction Unbiased

When a crystal of P-type and N-type is linked as shown in Fig. 4.7 the free electrons on the N-side diffuse across the junction, enters the P-side falls into the hole and free electron becomes a valance electron. Reverse happen from the P-side. A

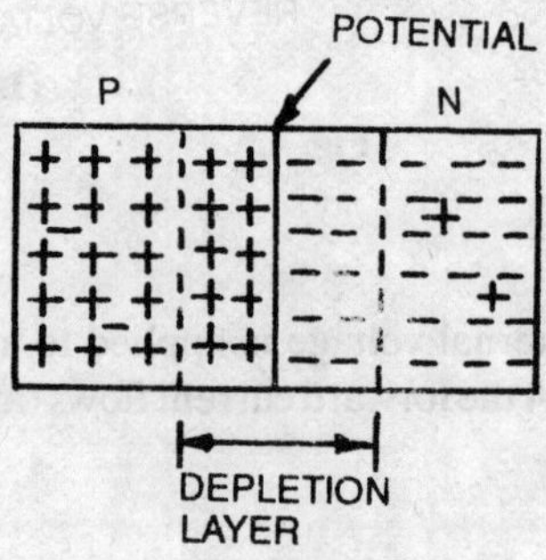

Fig. 4.7.

depletion layer is produced because of recombination of holes and electrons at the junction. This depletion layer develops a potential called as the barrier potential. At room temperature the barrier potential approximately equals 0.3 V for germanium diode and 0.7 V for silicon diode.

PN Junction Forward Bias

When a PN-junction is connected to a battery in such a way that P-side is connected to positive terminal of the battery and Negative terminal to N-side. This connection is known as

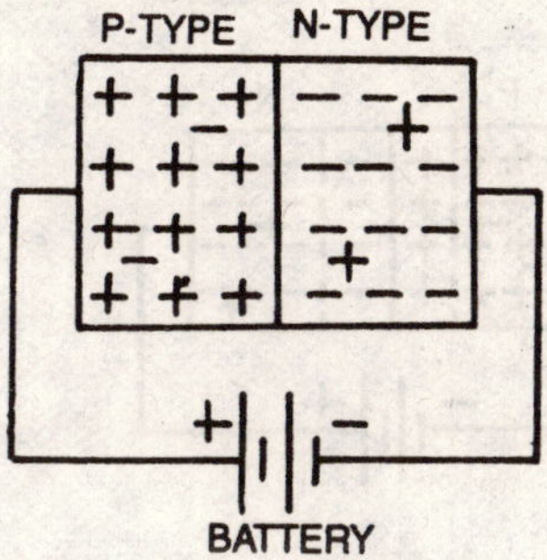

Fig. 4.8.

forward bias. Fig. 4.8 shows PN-Junction forward bias. Battery forces the free electrons and holes towards the junction. Free electron leave the negative of the battery and flow into the right end of the crystal. At the junction it recombines with a hole and becomes a valence electron. These electron travels through the P-side as a valence electron. If the battery voltage is greater than potential barrier, current flows easily.

PN Junction Reverse Bias

When a PN- junction is connected to a battery in such a way that P-side is connected to negative terminal of the battery and positive terminal to N-side. This connection is called reverse bias. Fig. 4.9 shows PN-junction reverse bias. The negative terminal of battery attracts the holes and positive terminal attracts the free electrons. The electrons and holes flow away from the junction. The depletion layer gets wider. The barrier potential increases. When barrier potential equals the battery voltage the depletion layer stops growing. There exists small reverse current this is because of minority carrier. The

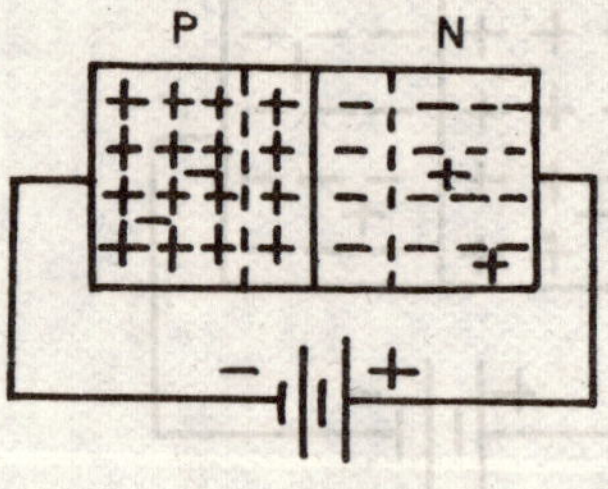

Fig. 4.9.

reverse current is approximately zero. PN-junction diode acts as a open switch in reverse bias. If we keep increasing reverse voltage a stage will come when diode is destroyed. This voltage is called breakdown voltage. At breakdown voltage large number of minority carriers exists in the depletion layer and it conducts heavily. This is called the avalanche effect. Also when a PN-junction is heavily doped and reverse voltage increases the electric field across the depletion layer, it pulls the electrons out of valence electrons. This effect is called zener effect. The device called as zener diode. These are used widely to regulated voltage.

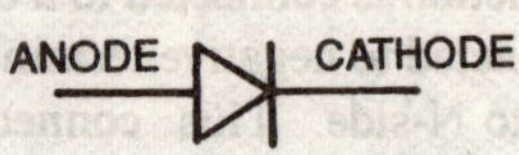

Fig. 4.10.

PN-junction is a diode. The P-side is anode and N-side cathode. Fig. 4.10 shows the symbol. When it convert alternating current to direct current it is called a rectifier diode Fig. 4.11

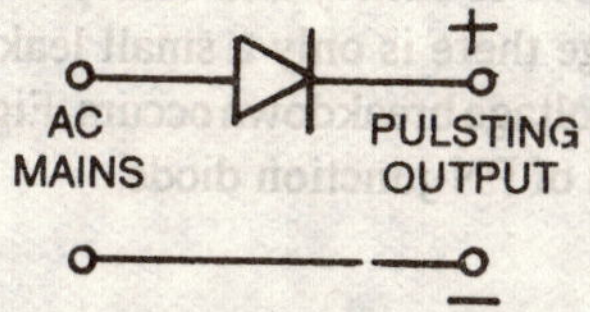

Fig. 4.11.

shows power supplies circuit converting a.c. mains into d.c. During positive half cycle diode is forward bias a.c. conducts. During negative half cycle diode is reverse bias it conducts poorly.

V-I Characteristics PN Junction Diode

To Plot the V-I characteristic connect the diode as shown in Fig. 4.12 When positive of battery is connected to anode and negative to cathode diode is forward bias. Increase the voltage at each step notedown the current. The diode does not conduct until the applied voltage overcomes the barrier potential at which current starts increasing sharply. This is called the knee voltage. For silicon diode knee voltage is 0.7 V and for germanium diode it is about 0.3 V.

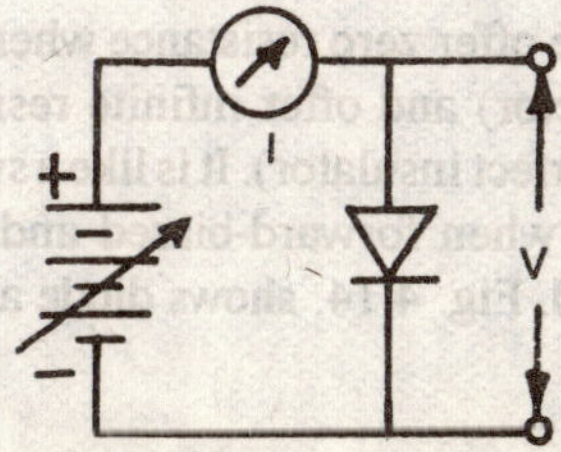

Fig. 4.12

When the diode is reverse bias, i.e., anode of diode is connected to negative of battery and cathode to positive. As we increase the voltage there is only a small leakage current. For very high reverse voltage breakdown occurs. Fig. 4.13 shows the V-I characteristics of PN-junction diode.

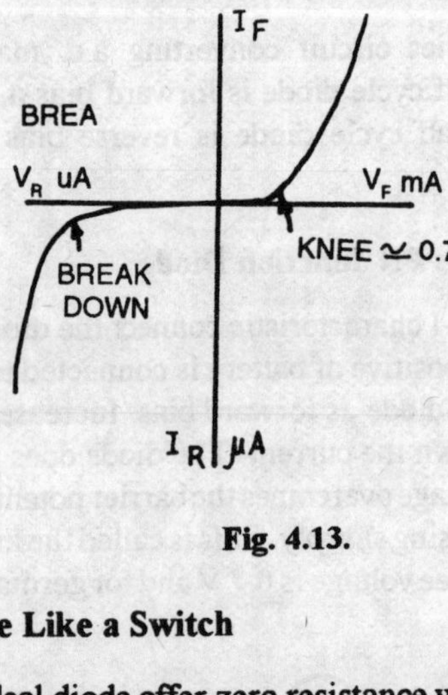

Fig. 4.13.

Ideal Diode Like a Switch

An ideal diode offer zero resistance when forward-biased (like perfect conductor) and offer infinite resistance when reverse biased (like perfect insulator). It is like a switch. Diode acts like a close switch when forward-biased and an open switch when reverse-biased. Fig. 4.14. shows diode as a switch.

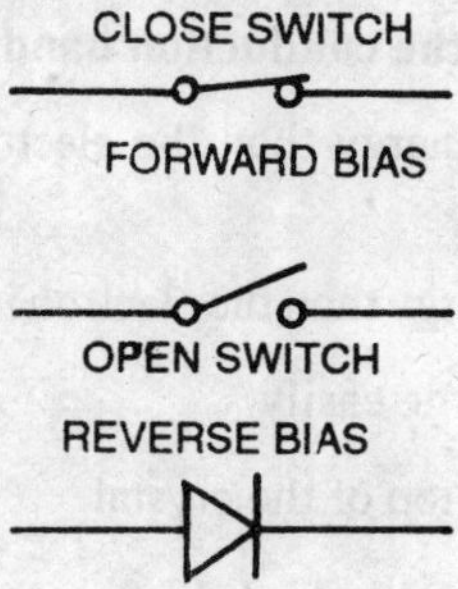

Fig. 4.14.

Junction Breakdown

As can be seen from V-I characteristics, the reverse current is constant for a very small value until V_R reaches the junction breakdown voltage. Thereafter a relatively large reverse current can flow. Most important, the reverse voltage across the junction is constant at the breakdown value. This effect is used for voltage regulator diodes.

For transistor amplifiers and rectifier diodes, forward characteristics are used.

The diode characteristic can be used in half wave or full wave rectifiers.

OBJECTIVE TYPE QUESTIONS

Q. 1. At absolute zero temperature a semi-conductor behaves as

(A) An insulator (B) A super-conductor

(C) A good conductor (D) A variable resistor

Q. 2. An electron in the conduction band

(A) Has higher energy than the electron in the valence band

(B) Has lower energy than the electron in the valence band

(C) Loses its charge easily

(D) Jumps to the top of the crystal.

Q. 3. E_G for silicon is 1.12 eV and for germanium is 0.72 eV Thus it can be concluded that

(A) More number of electron-hole pairs will be generated in silicon than in germanium at room temperature

(B) Less number of electron hole pairs will be generated in silicon than in germanium at room temperature

(C) Equal number of electron-hole pairs will be generated in both at lower temperatures

(D) Equal number of electron-hole pairs will be generated in both at higher temperatures.

Q. 4. E_G for silicon is 1.12 eV and for germanium is 0.72 eV. Thus, it can be concluded that

(A) The conductivity of silicon will be less than that of germanium at room temperature

(B) The conductivity of silicon will be more than that of germanium at room temperature

(C) The conductivity of two will be same at 6°C

(D) The conductivity of two will be same at 100°C.

Q. 5. Before doping the semiconductor material is generally

(A) Dehydrated (B) Heated

(C) Hardened (D) Purified.

Q. 6. Select the one that is acceptor impurity element ?

(A) Antimony (B) Gallium

(C) Arsenic (D) Phosphorous.

Q. 7. At room temperature when a voltage is applied to an intrinsic semi-conductor.

(A) Most of the electrons and holes move towards negative terminal

(B) Most of the electrons and holes move towards positive terminal

(C) Electrons move towards positive terminal and holes towards negative terminals

(D) Electrons move towards negative terminal and holes towards positive terminal.

Q. 8. What happens to conductivity when a semiconductor is doped ?

(A) Increases

(B) Decreases in the direct ratio of the doped material

(C) Decreases in the inverse ratio of the doped material

(D) Remains unaltered.

Q. 9. Under which of the following conditions avalanche breakdown in a semiconductor diode takes place ?

(A) When potential barrier is reduced to zero

(B) When reverse bias exceeds a certain value

(C) When forward dias exceeds a certain value

(D) When forward current exceeds a certain value.

Q. 10. Select the rectifier that needs four diodes

(A) Half wave rectifier

(B) Centre-tap full wave rectifier

(C) Bridge rectifier

(D) None of the above.

Q. 11. Maximum rectification efficiency for a half wave rectifier is equal to

(A) 100% (B) 88%

(C) 50% (D) 40.6%.

Q. 12. The maximum forward current in case of signal diode is in the range of

(A) 1A to 10A (B) 0.1A to 1A

(C) Few milli amperes (D) Few nano amperes.

Q. 13. In a semiconductor avalanche breakdown takes place when

(A) Reverse bias exceeds the limiting value

(B) Forward bias exceeds the limiting value

(C) Forward current exceeds the limiting value

(D) Potential barrier is reduced to zero.

Q. 14. The d.c. output voltage from a power supply is

(A) Increases with higher values of filter capcitance and decreases with more load current

(B) Decreases with higher values of filter capacitance and increases with more load current

(C) Decreases with higher values of filter capacitance as well as with more load current

(D) Increases with higher values of filter capacitance as well as with more load current.

Q. 15. With an a.c. input from 15 Hz power line, the ripple frequency is

(A) 50 Hz in the d.c. output of half wave as well as full wave rectifier

(B) 100 Hz in the d.c. output of half wave as well as full wave rectifier

(C) 50 Hz in the d.c. output of half wave and 100 Hz in the d.c. output of full wave rectifier

(D) 100 Hz in the d.c. output of half wave and 50 Hz in the d.c. output of full wave rectifier.

Q. 16. A solar cell provides an example of

(A) Photo voltaic cell (B) Photo conductive cell

(C) Photo emissive cell (D) Photo radiation cell.

Q. 17. When yellow light is incident on a surface, no electrons are emitted while green light can emit. If red light is incident on the surface, then it is expected that

(A) no electrons are emitted

(B) photons are emitted

(C) electrons of higher energy are emitted

(D) electrons of lower energy are emitted.

Q. 18 An ideal diode

(A) Should have zero resistance in the forward bias as well as reverse bias

(B) Should have zero resistance in the forward bias and an infinitely large resistance in reverse bias

(C) Should have infinitely large resistance in the forward bias and zero resistance in reverse bias

(D) Should have infinitely large resistance in forward as well as reverse bias.

Q. 19. In case a PN-junction is forward biased

(A) Holes and electrons seize to move

(B) Electrons and holes move away from the junction

(C) Electrons and holes move towards the junction

(D) Depletion region decreases.

Q. 20. The reverse resistance of a PN-junction diode

(A) is always low

(B) is always high

(C) is given by $\dfrac{\text{breakdown voltage}}{\text{reverse leakage current}}$

(D) is given by $\dfrac{\text{Forward voltage}}{\text{reverse leakage current}}$

Q. 21. For a PN-junction we generally have

I. Width of depletion layer

II. Junction barrier voltage

III. Reverse leakage current

Which of the above parameters will decrease when the temperature of the junction rises ?

(A) I only (B) I and II only

(C) II and III only (D) I, II and III.

Q. 22. In which case the temperature coefficient is positive

(A) Intrinsic semi-conductor

(B) Extrinsic semi-conductor

(C) Both intrinsic as well as extrinsic semi-conductor

(D) Neither intrinsic nor extrinsic semi-conductor.

Q. 23. The voltage waveform resulting from a three phase half-wave rectifier is shown in Fig. 4.15.

The average value of voltage is given by

(A) 89.9 V (B) 82.7 V

(C) 71.1 V (D) 63.3 V.

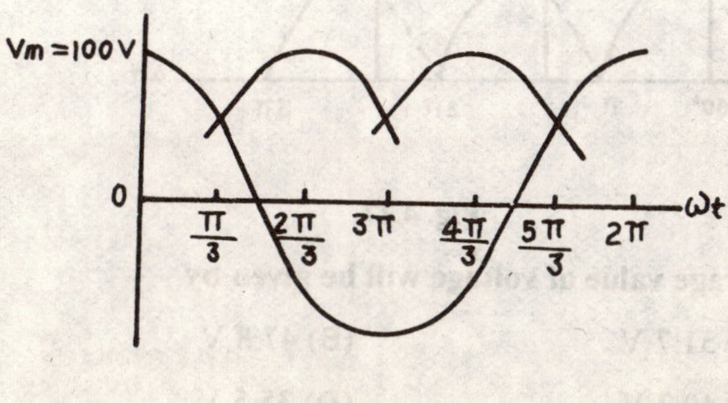

Fig. 4.15.

Q. 24. A delayed half-wave rectified sine wave of voltage is shown in Fig. 4.16. The delay angle is 45°.

The value of V_{av} will be equal to

(A) 66.6 V (B) 54.4 V

(C) 47.7 V (D) 27.2 V

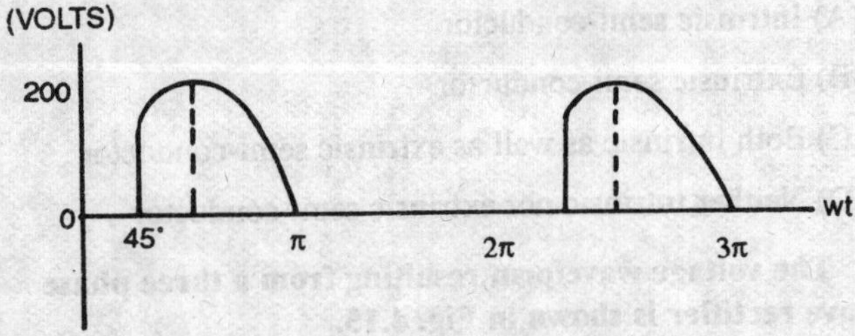

Fig. 4.16.

Q. 25. Full wave rectified sine wave shown in Fig. 4.17 has a delay angle of 60°.

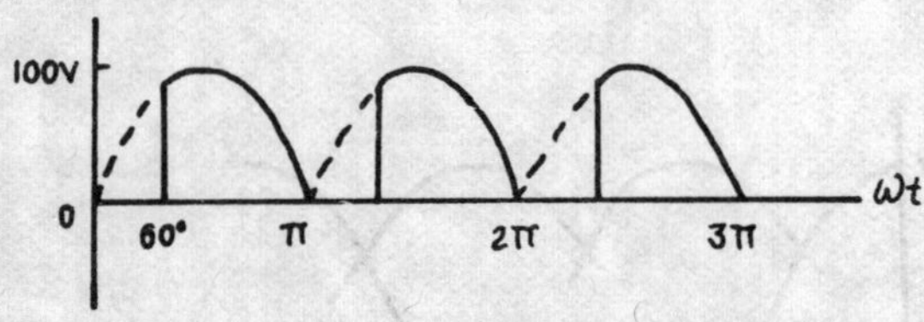

Fig. 4.17.

The average value of voltage will be given by

(A) 51.7 V (B) 47.8 V

(C) 40.7 V (D) 35.5 V.

ANSWERS

1. A,	2. A,	3. B,	4. A,	5. D,	6. B,
7. C,	8. A,	9. B,	10. C,	11. D,	12. C,
13. A,	14. A,	15. C,	16. A,	17. A,	18. B,
19. D,	20. C,	21. B,	22. B,	23. B,	24. B,
25. B.					

Chapter 5

Rectifiers

STUDY MATERIAL

Half-Wave Rectifier

The half-wave rectifier is constructed by using vacuum diode or solid state diode. Fig. 5.1 shows the circuit of half-wave rectifier.

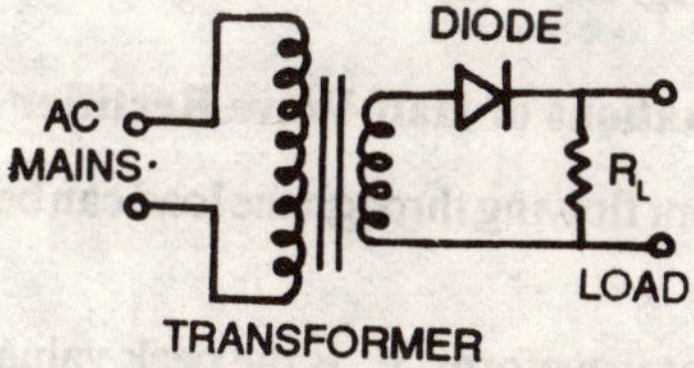

Fig. 5.1.

The diode is connected in series with the secondary of the transformer and the load R_L. The primary of transformer is connected to a.c. mains. The secondary voltage that is applied to diode may be same, more or less depending upon type of transformer. Fig. 5.2 shows the waveform of a.c. available at the

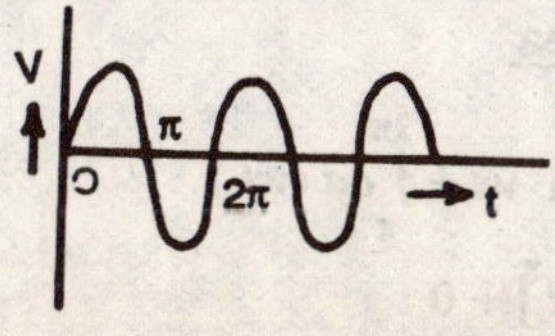

Fig. 5.2. Input waveform.

secondary. During positive half cycle of this voltage the diode is forward biased. The diode conducts and current flows through the load resistance. The voltage appears across the load. During negative half cycle, the diode is reverse biased. The diode does not conduct. No current flows through the load. Therefore no voltage appears across the load. Fig. 5.3 shows the output waveform. The output is not perfect d.c. it is unidirectional.

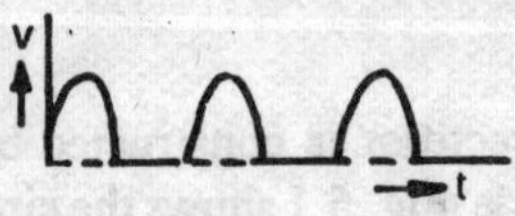

Fig. 5.3. Output waveform

Mathematical Equations of Half-Wave Rectifier

The d.c. current flowing through the load can be calculated as follows:

Refer to output waveform, V_m is the peak value of voltage then peak value of the current is given by $I_m = \frac{V_m}{RL}$ taking diode resistance $rd = 0$. The d.c. or average value of current can be found by first integrating the area of one complete cycle over the period 0 to 2 π.

Then for $0 \leq wt \leq \pi$ $I_o = I_m \sin wt$ and

For $\pi \leq wt \leq 2\pi$ $I_o = 0$ then

$$I_{dc} = I_{avg} = \frac{\text{Area}}{2\pi}$$

$$\text{Then Area} = \int_0^{\pi} I_m \sin wt + \int_{\pi}^{2\pi} od\,(wt)$$

$$= I_m \left[-\cos wt\right]_0^{\pi} + 0$$

$$= I_m [(-\cos\pi - (-\cos 0)] = 2I_m$$

$$I_{dc} = \frac{2I_m}{2\pi} = \frac{I_m}{\pi}$$

The d.c. voltage developed across load is equal to

$$V_{dc} = I_{dc} \times R_L$$

$$= \frac{I_m}{\pi} \times R_L$$ if rd is not neglected then

$$I_m = \frac{V_m}{(R_L + rd)} \quad V_{dc} = \frac{V_m R_L}{\pi (R_L + R_d)}$$

$$= \frac{V_m}{\pi\left(1 + \frac{rd}{R_L}\right)}$$ if $rd << R_L$.

Then $V_{dc} \approx \frac{V_m}{\pi}$

Then d.c. output power $P_{dc} = I^2dc \times R_L$

$$= \left(\frac{I_m}{\pi}\right)^2 \times R_L$$

$$= \frac{I_m^2 R_L}{\pi^2}$$

The a.c. or R.M.S. value of current is given by

$$I_{rms} = \sqrt{\frac{1}{2}\pi \int_0^{2\pi} I_m^2\, d\,wt}$$

$$= \sqrt{\frac{1}{2}\pi \{[\int_0^{\pi} I^2_m \sin^2 wtd(wt) + \int_0^{2\pi} od(wt)]}$$

refer output waveform

$$= \sqrt{\frac{I_m^2}{2\pi} \int_0^{\pi} \frac{1 - \cos 2\,wt}{2}\, d\,(wt)}$$

$$= \sqrt{\frac{I_m^2}{2\pi \times 2} \left| wt - \frac{\sin 2wt}{2} \right|_0^{\pi}}$$

$$= \frac{I_m}{2}.$$

The output of rectifier is not only a d.c. output but also contains a.c. component. The ratio of r.m.s. value of the alternating component in the load to the d.c. or average value is called the ripple factor or

$$\text{R.F.} = \frac{\text{r.m.s. value of the alternating component in the load}}{\text{d.c. or average value of component in the load}}$$

The a.c. or r.m.s. value of current calculated above $\left(\frac{I_m}{2}\right)$ also contains r.m.s. value of d.c. Then r.m.s. value of alternating component of current is equal to $I_{rms}(ac) = \sqrt{I_{rms}^2 - I_{dc}^2}$

$$\text{Thus R.F.} = \frac{I_{rms}\,(ac)}{I_{dc}} = \frac{\sqrt{I_{rms}^2 - I_{dc}^2}}{I_{dc}}$$

$$\text{or R.F.} = \sqrt{\left(\frac{I_{rms}}{I_{dc}}\right)^2 - 1}$$

In half-wave rectifier $I_{rms} = \frac{I_m}{2}$ and $I_{dc} = \frac{I_m}{\pi}$

$$\text{Then R.F.} = \sqrt{\left(\frac{I_m/2}{I_m/\pi}\right)^2 - 1} = \sqrt{\left(\frac{\pi}{2}\right)^2 - 1}$$

$$= 1.21$$

Since the *rectification efficiency* of a rectifier is the ratio of the d.c. output power to the a.c. power supplied to the rectifier

$$\eta = \frac{\text{d.c. output power}}{\text{a.c. input power supplied to the rectifier}}$$

$$= \frac{P_{dc}}{P_{dc}}$$

for a half-wave rectifier the $P_{dc} = I_{dc}^{\,2} R_L$

$$= \left(\frac{I_m}{\pi}\right)^2 R_L$$

and the P_{ac} is $= I_{rms}^{\,2} (rd + R_L)$

$$= \left(\frac{I_m}{2}\right)^2 (rd + R_L)$$

$$\text{Therefore } \eta = \frac{I_m^{\,2} R_L . 4}{\pi^2 . I_m^{\,2} (rd + R_L)}$$

$$= \frac{4}{\pi 2} . \frac{R_L}{rd + R_L}$$

$$= \frac{40.6}{1 + \frac{rd}{R_L}} \text{ \% if } rd << R_L.$$

Then rectification efficiency approximately equals 40.6%

Full-wave Rectifier

A full-wave rectifier converts the applied alternating voltage into a pulsating (unidirectional) voltage using the full-

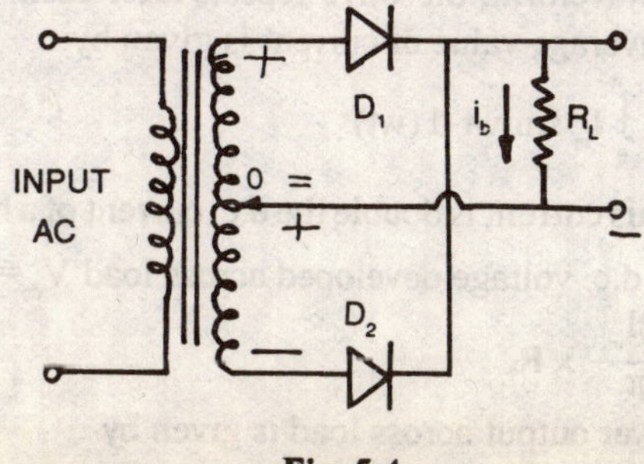

Fig. 5.4.

cycle of the applied a.c. voltage. Two types of full-wave rectifier circuits are constructed one is centre-tap rectifier that uses two diodes. The other is bridge rectifier using for diodes. The circuit of full-wave rectifier using two diodes is shown in Fig. 5.4. The circuit uses centre- tap secondary transformer. The two diodes are connected in such a way that during each half cycle one diode is forward biased other is reverse biased. As seen from the Fig. 5.4. Centre-tap of transformer is the reference point. The diode D_1 conducts during the positive half cycle and current flows through load such that upper terminal is positive. The diode D_2 conducts during negative half cycle, the current flows through the load again upper terminal is positive. The diode conducts during alternate cycle but each time, the current flow in the same direction. The output current waveform is shown in Fig. 5.5.

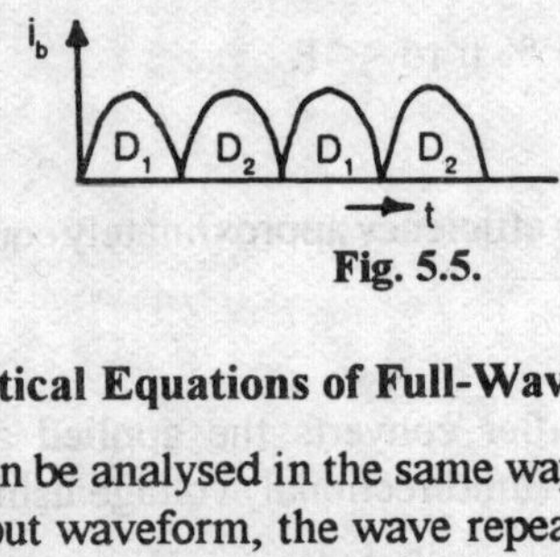

Fig. 5.5.

Mathematical Equations of Full-Wave Rectifier

It can be analysed in the same way as half-wave rectifier. Refer output waveform, the wave repeats after each π period. Then d.c. or average value of current is given by

$$I_{dc} = I_{aug} = \frac{1}{\pi}\int_0^{\pi} I_m \sin wt \, d(wt)$$

$$= \frac{2I_m}{\pi}$$, this current is double the d.c. current of a half- wave rectifier. The d.c. voltage developed across load $V_{dc} = I_{dc} \times R_L$

$$= \frac{2I_m}{\pi} \times R_L$$

d.c. power output across load is given by

$$P_{dc} = I^2_{dc} \times R_L$$

$$= \left(\frac{2I_m}{\pi}\right)^2 \times R_L$$

$$= \frac{4I_m^{\ 2}}{\pi^2} \ R_L$$

The a.c. or R.M.S. value of current is given by

$$I_{rms} = \frac{1}{\pi} \int_0^{\pi} I_m^{\ 2} \sin^2 wt \ d(wt)$$

$$= \frac{I^2_m}{\pi} \int_0^{\pi} \left(\frac{1 - \cos 2wt}{2}\right) d(wt)$$

$$= \sqrt{\frac{I_m^{\ 2}}{\pi} \times \frac{\pi}{2}} = \frac{I_m}{\sqrt{2}}$$

The a.c. power P_{ac} is given by

$$P_{ac} = \left(\frac{I_m}{\sqrt{2}}\right)^2 (rd + R_L)$$

$$\text{Therefore } \eta = \frac{\text{d.c. output power}}{\text{a.c. input power}}$$

$$= \frac{\left(\frac{2I_m}{\pi}\right)^2 \times R_L}{\left(\frac{I_m}{\sqrt{2}}\right)^2 (rd + R_L)} = \left(\frac{2\sqrt{2}}{\pi}\right)^2 \cdot \frac{1}{1 + \frac{rd}{R_L}}$$

$$= \frac{0.812}{1 + \frac{rd}{R_L}} \text{ if } rd << R_L$$

Then *rectification efficiency* is approximately equal to 81.2 which is twice the half wave rectifier.

For full-wave rectifier, *ripple factor* is given by

$$R.F. = \frac{\text{r.m.s. value of the alternating component in the load}}{\text{d.c. or average value of component in the load}}$$

$$R.F. = \sqrt{\frac{I_{rms}^2 - I_{dc}^2}{I_{dc}^2}} = \sqrt{\left(\frac{I_{rms}}{I_{dc}}\right)^2 - 1}$$

$$\sqrt{\left(\frac{I_m/\sqrt{2}}{2I_m/\pi}\right)^2 - 1} = \sqrt{\left(\frac{\pi}{2/\sqrt{2}}\right)^2 - 1}$$

$$= 0.48.$$

Peak Inverse Voltage (PIV)

During the negative half-cycle diode is reverse biased, the whole of the reverse voltage appear across diode. The diode has to withstand this maximum voltage. This maximum voltage is called the peak inverse voltage. For half-wave rectifier PIV = V_m and for full wave rectifier PIV = $2V_m$.

Voltage Regulation

It is the measure of the ability of a rectifier to maintain a specified output with change in the output load. Mathematically it is change.

$$\text{Voltage Regulation} = \frac{E_{dc} \text{ at no load} — E_{dc} \text{ at full load}}{E_{dc} \text{ at full load}}$$

Bridge Rectifier

It is a most widely used full-wave rectifier. It uses four diodes. There is no need of centre-tap secondary transformer. The circuit is shown in Fig. 5.6. A.C. mains is applied through simple transformer or a.c. can be applied directly to rectifier. Full secondary voltage is applied to two diodes in series. During the positive half-cycle two diodes D_1 and D_2 becomes forward

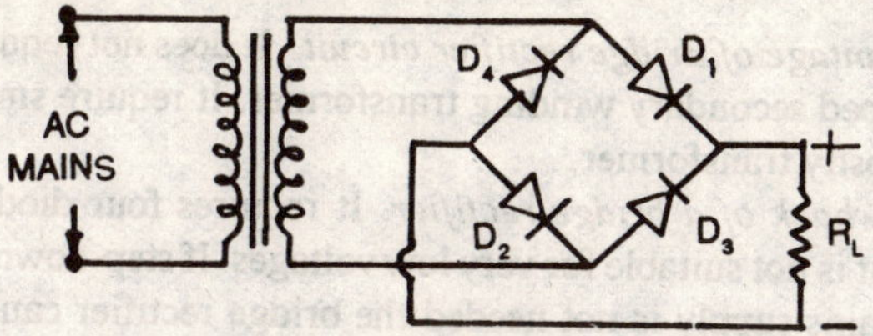

Fig. 5.6.

biased, current flows through these diodes load and back to secondary of transformer. Fig. 5.7 shows current flow during

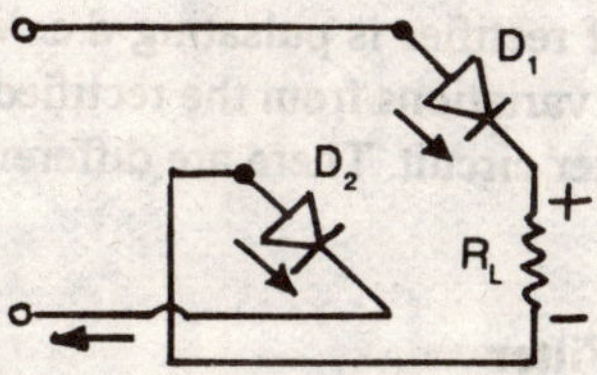

Fig. 5.7.

positive half-cycle. During the negative half-cycle reverse happens, diodes D_3 and D_4 becomes forward biased and D_1, D_2 reverse biased. Fig. 5.8 shows current flow during negative half-cycle. It is seen from the circuit that current always flow in load in same direction during positive and negative half-cycle.

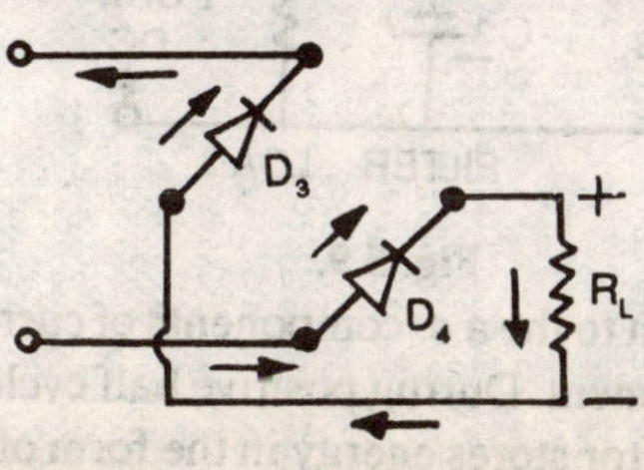

Fig. 5.8.

Advantage of Bridge rectifier circuit. It does not require centre-tapped secondary winding transformer. It require small and less costly transformer.

Drawback of a bridge rectifier. It requires four diodes. This circuit is not suitable for very low voltages. If step-down or step-up mains supply is not needed the bridge rectifier can be used directly across the mains supply. The circuit becomes cheaper and there is no need for the transformer.

Filters Circuit

The output of rectifier is pulsating d.c. We can filter or smooth out the a.c. variations from the rectified output voltage with the help of filter circuit. There are different types of filter circuits.

Shunt Capacitor Filter

The filter circuit is made up of a large valve capacitor, that is connected in shunt with the load resistor. This is the simplest and cheep type of filter circuit. A half-wave rectifier with shunt capacitor filter is shown in Fig.5.9. The Shunt capacitor offers

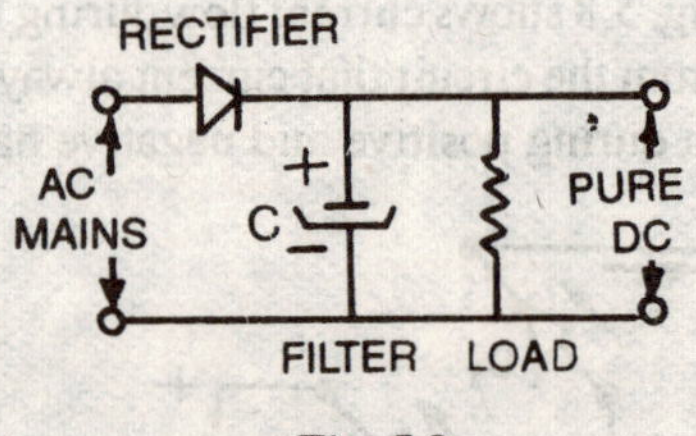

Fig. 5.9.

a low reactance Path to the a.c. components of current and open circuit to d.c. component. During positive half cycle when diode conducts and capacitor stores energy in the form of electrostatic field. During negative half - cycle, the filter capacitor releases

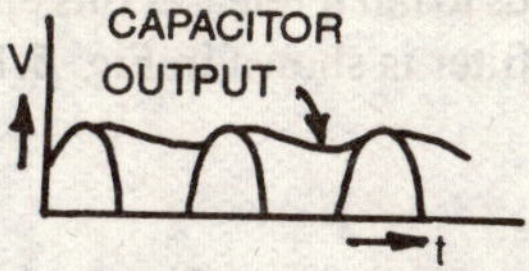

Fig. 5.10. Output Waveform

stored energy to the load. The output waveform is shown in Fig. 5.10.

Inductor Filter

The inductor, coil or choke offers low resistance of the winding to the passage of d.c. while offering at the same time a high impedance to the alternating current. A full-wave rectifier with inductor filter is shown in Fig. 5.11. Whenever rectifier

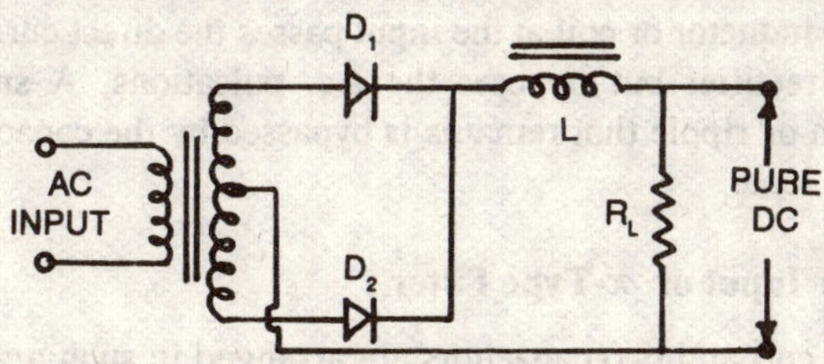

Fig. 5.11.

conducts current in the inductor or choke increases it opposes the change in the current by storing up energy in the magnetic field.

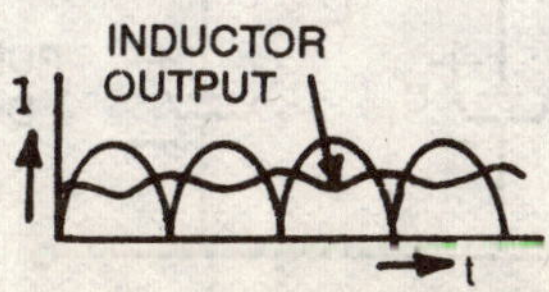

Fig. 5.12.

Whenever the current tends to fall it releases this energy. Output waveform with inductor filter is shown in Fig. 5.12.

Choke Input Filter

The choke input filter is shown in Fig. 5.13. When choke and capacitor filter are joined together, they form choke input

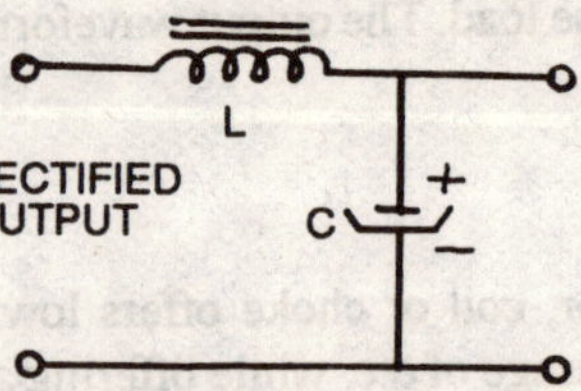

Fig. 5.13.

filter. The inductor or coil at the input passes the direct current from the rectifier but opposes the a.c. pulsations. A small fluctuation or ripple that remains is bypassed by the capacitor across the load.

Capacitor Input or π-Type Filter

The coil and two capacitors are arranged in such a way, that they form π-type structure. The capacitors input filter is

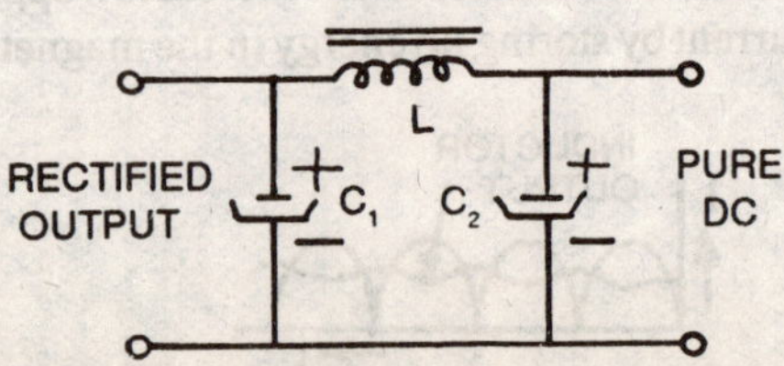

Fig. 5.14.

shown in Fig. 5.14. The input capacitor C_1, stores the charge when rectified output tends to rise, then discharges slowly-through the coil and load. The pulsation or ripple are largely reduced. The remaining fluctuations of the rectifier are opposed by the series choke and bypassed to ground by the capacitor C_2. The output waveform with π-type filter is shown in Fig. 5.15.

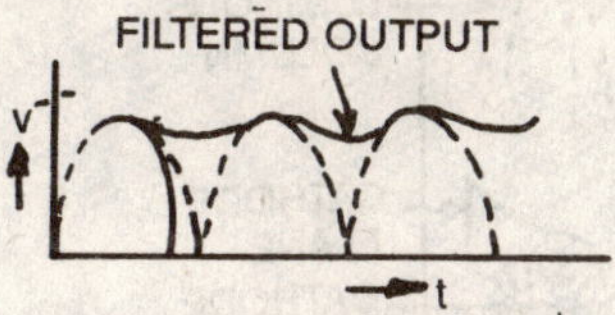

Fig. 5.15.

The output voltage of a π-type filter is higher than that of a inductor input filter for the same a.c. input. The voltage regulation of π-type filter is poorer than choke input filter.

Mercury Arc Rectifier

The mercury arc rectifier is made up of an evacuated glass vessel having a pool of mercury at the bottom (cathode) and graphite block (the anode) at the top. It is connected to circuit through wires sealed into the glass which serves as terminal. When the conduction starts, the cathode spot emits electrons and mercury vapour, the anode attracts the electrons; the electrons in their flight shock-ionise the molecules of mercury vapour; the enriched electron stream proceeds to the anode; the ionised mercury vapour molecules are driven downwards to the cathode spot. The upward electron stream and downward ion stream constitute an electric current passing from anode to cathode internally. The action of the rectifier needs the cathode spot to be produced first by some auxiliary device before the arc can be established. This is done by introducting arc between cathode and an auxiliary anode—the process is called ignition. Further the arc will be maintained only when the current is above certain level.

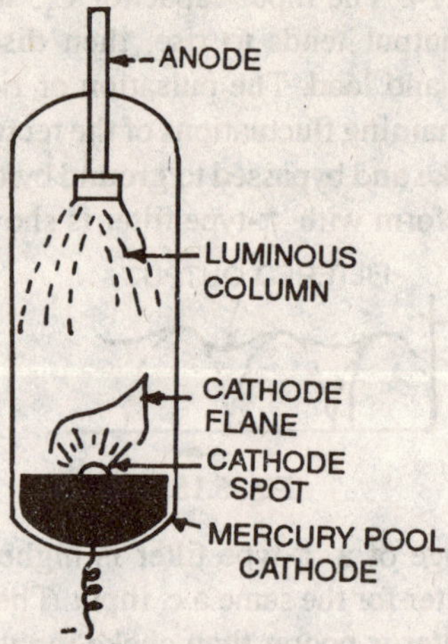

Fig. 5.16.

Total internal losses in rectifier are given by

losses = Voltage drop at anod

+ Voltage drop at cathode

+ Voltage drop in the arc

= 9V + 5V + (0.1L) V (approx.)

where L is the length of arc in cm.

Three Anode Rectifier

The arrangement of three anode rectifier is shown in Fig. 5.17. The three secondary windings are star connected and the star point forms the negative output terminal. The voltages of the three secondary phase are equal, but displaced in time by one-third of a period. As the conduction will take place between the cathode and that anode with the maximum positive voltage, consequently the arc will change from anode to anode, thus remaining with each anode for one third of the cycle period in the middle of its positive half cycle.

As shown in Fig. 5.17 the output waveform made up of the successive tops of the positive waves. The output waveform is still smooth when size anodes are used.

Rotary Convertors

The rotary convertors, also called synchronous convertors are rotating machines for converting an a.c. supply to d.c. or *vice versa*. Such machines have constructional features similar to d.c. machines with stationary field magnet system and a rotating armature carrying a double lap or wave winding. The winding is connected to a commutator upon which the brushes are connected to the d.c. terminals. Also at the opposite end of the

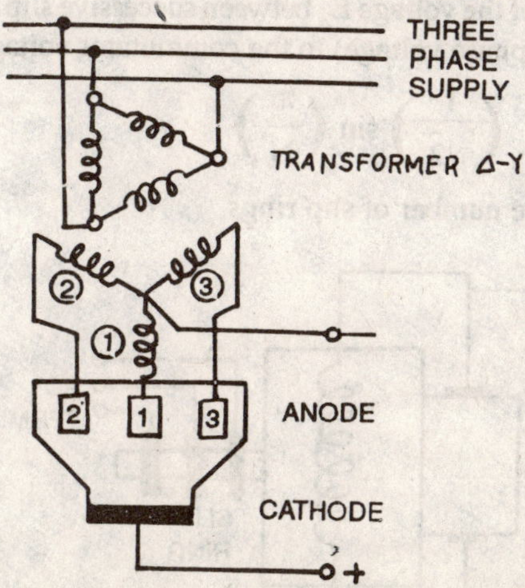

(A)

Fig. 5.17.

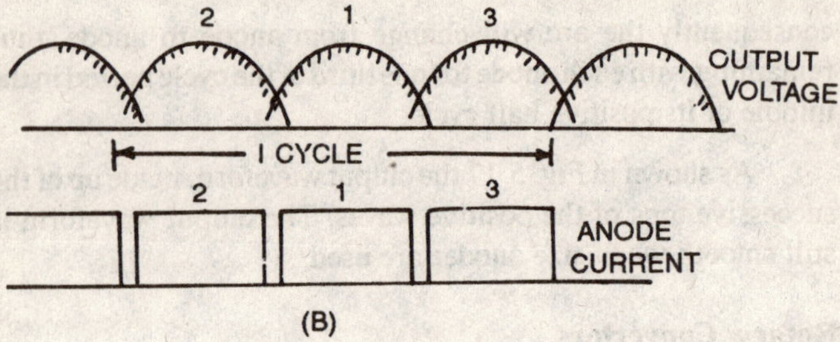

Fig. 5.18.

armature to the commutator are slip-rings connected to tapping in the armature winding and the brushes on these rings are connected to the a.c. terminals.

The supply on the a.c. side may be 1, 3, 6 or 12 phase. Accordingly the number of slip rings may be 2, 3, 6 or 12. The no load ratio of the voltage E_A between successive slip rings (i.e. the armature phase voltage) to the commutator voltage E_D is

$$\frac{E_A}{E_D} = \left(\frac{1}{\sqrt{2}}\right) \sin\left(\frac{\pi}{N}\right)$$

where N is the number of slip rings.

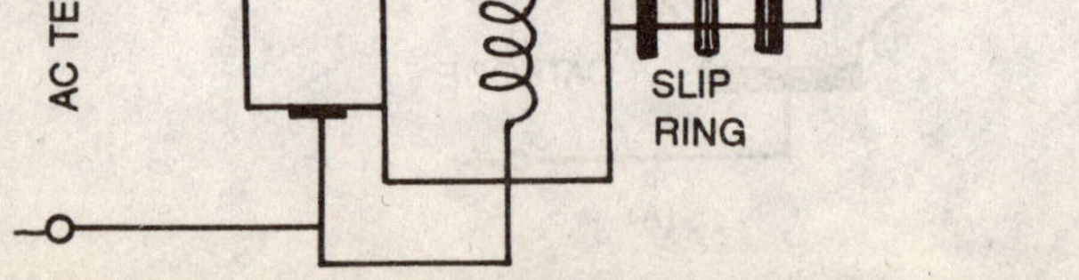

Fig. 5.19.

N	*ratio* $\frac{E_A}{E_D}$
2	0.707
3	0.61
6	0.35
12	0.18

The a.c. and d.c. armature powers will be equal when :

(i) losses are neglected

(ii) power factor is unity.

In that case the ratio of slip ring (line current) to the d.c. output current will be

$$\frac{I_L}{I_D} = (2\sqrt{2})/N$$

where N is the number of slip rings.

N	*ratio* $\frac{I_L}{I_D}$
2	1.41
3	0.94
6	0.47
12	0.24

For efficiency η and power factor $\cos\phi$, the ratio will be greater by a factor $\frac{1}{\eta \cos\phi}$.

OBJECTIVE TYPE QUESTIONS

Q. 1. If a synchronous converter is supplied with 12 phase a.c. supply, the number of slip rings will be

(A) 6 (B) 9

(C) 12 (D) 15.

Q. 2. In a rotary convertor I^2R losses as compared to a d.c. generator of the same size will be

(A) Six times more (B) Double

(C) Half (D) Less.

Q. 3. When a rotor converter is started by means of a small auxiliary motor, the power of motor, must be

(A) Less than the d.c. output of converter

(B) Less than a.c. input of converter

(C) Half of d.c. output of converter

(D) Slightly more than the value of friction and windage losses at rated speed.

Q. 4. Maximum current rating of a glass bulb mercury arc rectifier is usually restricted to

(A) 150 A (B) 400 A

(C) 500 A (D) 2000 A.

Q. 5. In an arc rectifier the drop in voltage at the cathode is approximately

(A) 10 Volt (B) 6 to 7 Volts

(C) 160 to 170 Volts (D) 1.2 $\sqrt{V}$ Volts.

Q. 6. In a mercury arc rectifier the cathode voltage drop is because of

(A) Surface resistance

(B) Expenditure of energy in liberating electrons from the mercury

(C) Expenditure of energy in ionization

(D) Expenditure of energy in overcoming the electrostatic field.

Q. 7. The average life of the glass bulb rectifier is

(A) 80 to 100 hours

(B) 2000 to 2500 hours

(C) 10,000 to 15,000 hours

(D) 200,000 to 250,000 hours.

Q. 8. The vacuum inside the glass bulb of a mercury arc rectifier is of the order of

(A) 5×10^{-3} cm of Hg (B) 5×10^{-4} cm of Hg

(C) 5×10^{-5} cm of Hg (D) 5×10^{-6} cm of Hg.

Q. 9. A 3 anode mercury arc rectifier has an anode current of overlap 30°. Neglecting arc drop, the regulation will be approximately

(A) 4% (B) 5%

(C) 7% (D) 9%.

Q. 10. Which of the following is the loss within the mercury arc rectifier chamber ?

(A) Voltage drop at the cathode

(B) Voltage drop at the anode

(C) Voltage drop in arc

(D) All of the above.

Q. 11. As the output voltage of a single anode mercury arc rectifier increases, the variation of internal efficiency is shown by

(A) Curve A (B) Curve C

(C) Curve B (D) Curve D.

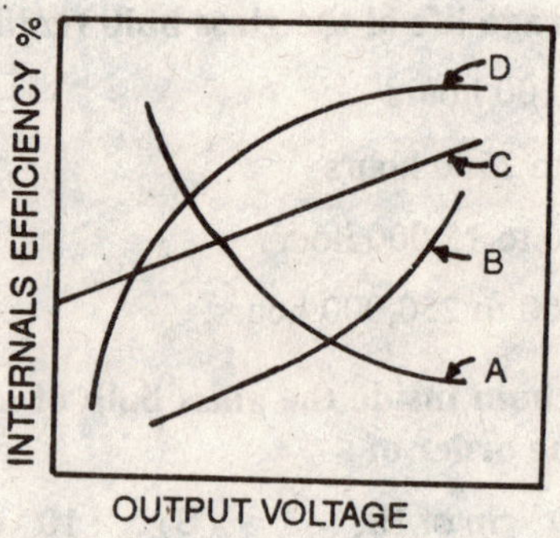

Fig. 5.20.

Q. 12. In a mercury arc rectifier

(A) Ion stream moves from cathode to anode

(B) Current flows from cathode to anode

(C) Electron stream moves from anode to cathode

(D) Ion stream moves from anode to cathode.

Q. 13. For producing cathode spot in a mercury arc rectifier

(A) An auxiliary electrode is used

(B) Tube is evacuated

(C) Low mercury vapour pressures are used

(C) Anode is heated.

Q. 14. If the voltage of anode B is raised to 510 V

(A) Anode B will conduct but anode A will not conduct

(B) Anode B will conduct but anode A will also continue to conduct

(C) Anode B will not conduct and anode A will continue to conduct

(D) Both anodes will not conduct.

Q. 15. Ripple frequency of full wave rectifier working on 50 Hz supply will be

(A) 25 Hz

(B) 150 Hz

(C) 6 anode rectifier with inter phase transformer

(D) All will have identical power factor.

Q. 16. The rms value of half wave rectified sine wave with i_m as peak value is

(A) 0.700 i_m (B) 0.665 i_m

(C) 0.50 i_m (D) 0.310 i_m.

Q. 17. The form factor for half-wave rectified sine wave is

(A) 1.05 (B) 1.15

(C) 1.45 (D) 1.57.

Q. 18. For full-wave rectified sine wave, rms value is

(A) 0.700 i_m (B) 0.6036 i_m

(C) 0.5 i_m (D) 0.318 i_m.

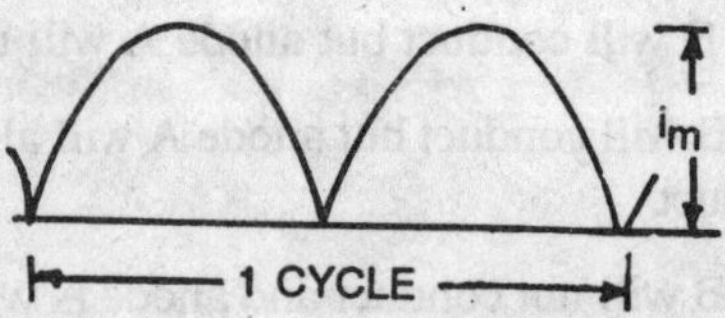

Fig. 5.21.

Q. 19. For full-wave rectified sine wave, form factor is

(A) 1.55 (B) 1.44

(C) 1.22 (D) 1.11

Q. 20. A silicon controlled rectifier is a

(A) Unijunction device

(B) Device with three junction

(C) Device with four junctions

(D) None of the above.

Q. 21. For single phase supply frequency of 5 Hz, ripple frequency in full wave rectifier is

(A) 250 (B) 500

(C) 100 (D) 200.

ANSWERS

1. C,	2. D,	3. D,	4. C,	5. B,	6. B,
7. C,	8. C,	9. C,	10. D,	11. D,	12. D,
13. A,	14. A,	15. A,	16. C,	17. D,	18. B,
19. D,	20. B,	21. C.			

Chapter 6

Electromagnetic Theory

STUDY MATERIAL

Coulomb's Law

It states that the force between two very small objects separated in a vacuum or free space by a distance which is large compared to their size is proportional to the charge on each and inversely proportional to the square of the distance between them. Mathematically,

$$F = k \frac{Q_1 Q_2}{R^2}$$

Q_1 and Q_2 are the positive or negative quantities of charge, R is the separation, and k is a proportionality constant. In SI system Q is measured in coulombs (C), R in metres and the force in newtons (N). This is achieved when the proportionality constant k is written as

$$k = \frac{1}{4\pi\varepsilon_0}$$

The new constant ε_0 is known as the *permittivity of the free space* and has the magnitude, measured in farads per metre (F/m),

$$\varepsilon_0 = 8.854 \times 10^{-12} = \frac{1}{36\pi} \times 10^{-9} \text{ F/m.}$$

The quantity is not dimensionless, for Coulomb's law

shows that it has the label $c^2/N\text{-}m^2$. Coulomb's law then can be written an under:

$$F = \frac{Q_1 Q_2}{4\pi\varepsilon_0 R^2}$$

Electric Field Intensity

It is the vector force on a unit positive test charge. Electric field intensity must be measured by the unit—Newtons per coulomb, force per unit charge. E, electric field intensity is given by

$$E = \frac{F_t}{Q_t}$$

$$E = \frac{Q}{4\pi\varepsilon_0 R^2} a_R$$

R is the magnitude of the vector R the directed line segment from the point where the point charge Q is located to the point at which E is desired, and a_R is a unit vector in the R direction.

Gauss's Law

The electric flux passing through any closed surface is equal to the total charge enclosed. For the electric flux by ψ and the electric flux density by D, we have,

$$\psi = \oint_{\substack{\text{closed}\\\text{surface}}} D_t . dS = Q$$

Maxwell's First Equation (electrostatics) div. D = ρ

The equation applies to electrostatics and steady magnetic fields. According to it electric flux per unit volume leaving a vanishingly small volume unit is exactly equal to the volume charge density there. This equation is also known as the point form of Gauss's law.

No work is done in carrying the unit charge around any closed path, or

$$\oint E \cdot dL = 0$$

Dipole

It is the name given to two point charges of equal magnitude and opposite sign, separated by a distance which is small compared to the distance to the point where we want to know the electric and potential fields.

Current and Current Density

Electric charges in motion constitute a current. Current density is a vector and is represented by J. So

$$I = \int_s J.dS$$

Also $$\nabla \cdot J = -\frac{\partial \rho}{\partial t}$$

Capacitance

For a two conductor system it is the ratio of the magnitudes of the total charge on either conductor to the potential difference between conductors. Thus

$$C = \frac{Q}{V_0} = \frac{\varepsilon S}{d}$$

Total energy stored in a capacitor is given by

$$W_E = \frac{1}{2}\int_{vol} \varepsilon E^2\, dv = \frac{1}{2}\int_0^S \int_0^d dzds$$

$$= \frac{1}{2} CV^2 = \frac{1}{2} QV = \frac{Q^2}{2C}$$

Poisson's Equations

$$\nabla . \nabla V = -\frac{\rho}{\varepsilon} \text{ (for } \varepsilon \text{ constant)}$$

In cartesian coordinates

$$\nabla^2 V = \frac{\partial^2 V}{\partial x^2} + \frac{\partial^2 V}{\partial y^2} + \frac{\partial^2 V}{\partial z^2} = -\frac{\rho}{\varepsilon}$$

In the above equation if $\rho = 0$, indicating zero volume charge density but allowing point charges, line charge, and surface charge density to exist on the boundaries as sources of the field, then

$$\nabla^2 V = 0$$

which is Laplaces equation. Thus

$$\nabla^2 V = \frac{\partial^2 V}{\partial^2 t} + \frac{\partial^2 V}{\partial y^2} + \frac{\partial^2 V}{\partial z^2} = 0$$

in cartesian coordinates

$$\nabla^2 V = \frac{1}{r}\frac{\partial}{\partial r}\left(r\frac{\partial V}{\partial r}\right) + \frac{1}{r^2}\left(\frac{\partial^2 V}{\partial \phi^2}\right) + \frac{\partial^2 V}{\partial z^2}$$

(cylindrical)

$$\nabla^2 V = \frac{1}{r^2}\frac{\partial}{\partial r}\left(r^2\frac{\partial V}{\partial r}\right) + \frac{1}{r^2 \sin\theta}\frac{\partial}{\partial \theta}\left(\sin\theta\frac{\partial V}{\partial \theta}\right) + \frac{1}{r^2 \sin^2\theta}\frac{\partial^2 V}{\partial \phi^2}$$

(Spherical)

Steady Magnetic Field. (Biot Savart Law)

$$dH = \frac{IdL \times a_R}{4\pi R^2}$$

where H is magnetic field intensity (amperes per metre), dL is the differential vector length of the filament, R is the distance from the differential element to the point P.

The law of Biot-Savart is sometimes called Ampere's law for the current element.

Amperes Circuital Law

States that the line integral of H about any closed path is exactly equal to the current enclosed by that path

$$\oint H.\ dL = I$$

Stoke's Theorem

Relates a surface integral to a closed line integral

$$\oint H.dL = \int_S (\nabla \times H).\ dS$$

where dL is taken only on the perimeter of S.

Magnetic Flux and Magnetic Flux Density

In free space, magnetic flux B is given by

$$B = \mu_0 H$$

where B is measured in webers per square metre or in SI units is Tesla. In both cases μ_0 is not dimensionless and has the defined value for free space, in henrys per metre. Thus,

$\mu_0 = 4\pi \times 10^{-7}$ H/m and μ_0 is called the permeability of free space.

Magnetic Forces, Materials and Inductance

Force on a moving charge

$$F = QE$$

The force is in the same direction as the electric field intensity (for a positive charge).

The force on a moving particle due to combined electric and magnetic fields is

$$F = Q(E + U \times B)$$

where U is the velocity and B the flux density.

Also the current density $dF = J \times B \, dv$

where J is the current density.

Also it can be shown that $F = IL \times B$

The materials wherein permanent magnetic moment m_0 of each atom is zero, called diamagnetic materials. Copper, gold, silicon, germanium, graphite and sulphur are diamagnetic substances.

The materials that show no magnetic effects in the absence of an external field but however, when an external field is applied, there is a small torque on each atomic moment at these moments tend to become aligned with the external field, are called paramagnetic potassium, oxygen, tungsten, and the rare earth elements are the examples of paramagnetic substances.

In ferromagnetic materials each atom has a relatively large dipole moment. In a ferromagnetic materials, the forces between adjacent atoms cause the atomic moments to up in an anti-parallel fashion. The net magnetic moment is zero, and anti-ferromagnetic materials are affected only slightly by the presence of an external magnetic field.

Also $B = \mu H$

where the permeability μ is given by

$$\mu = \mu_0 \mu_R$$

where μ_R is relative permeability

Inductance L is given by

$$L = \frac{N\phi}{I}$$

where I is the current flowing in N turn coil producing a total flux ϕ.

Gauss's Law

Total displacement or electric flux through any closed surface surrounding charges is equal to the amount of charge enclosed. Mathematically,

$$\text{div}\,\vec{I} = \rho$$

In light of above, the law can be stated as—at every point in a medium divergence of electric displacement density is equal to the charge density. Gauss's law is very useful in determining field and potential distribution about bodies having symmetrical geometry.

Gauss's Theorem

The surface integral of normal component of P over the surface equal the volume integral of divergence P over the volume integral v. That is

$$\iint_S \vec{P}\,.\,\vec{n}\,da = \iiint_V \text{div}\,\vec{P}\,dv$$

where $\vec{n}$ is the unit vector along the outer normal.

Strokes Theorem

The line integral of the tangential component of a vector $\vec{P}$ around a closed path L is equal to the surface integral of the normal component of curl $\vec{P}$ over the surface S, enclosed by the path L. Thus,

$$\int_L \vec{P} \; d\vec{s} = \iint (\text{curl } \vec{P}) \, \vec{n} \, ds$$

Helmholtz Theorem

The theorem states that the vector field is completely specified by its divergence and curl.

Electrostatic Field

The sources of electromagnetic fields are charges, and the strength of field at any point depends on the magnitude, position, velocity and acceleration of the charges involved. An electrostatic field is a special case of electromagnetic field in which the charges, i.e., the sources are stationary. This helps in considering only the magnitude and position of charges as velocity and acceleration can be neglected if the charges are stationary.

Coulomb's Law

Force F between two charges Q_1 and Q_2 is proportional to the product of the charges and inversely proportional to the square of the distance between them. Mathematically,

$$F = K \frac{Q_1 Q_2}{r^2} \text{ Newtons}$$

Equipotential Surface

An equipotential surface is that surface on which the potential everywhere is the same.

The relationship between electric intensity E and potential V is expressed as

$$\vec{E} = -\text{grad } V$$

Thus, the electric intensity at any point is the negative of the potential gradient at that point. The direction of the electric field is the direction in which the gradient is greatest or in which the potential changes most rapidly.

Electrostatic Energy in a Capacitor

The capacitor stores an electrostatic energy equal to work required to built up the charge on the conductors. The electrostatic energy in a charged capacitor

$$= \frac{1}{2}\frac{Q^2}{C} = \frac{1}{2}V^2Q = \frac{1}{2}V^2C \text{ Joules.}$$

Current density is defined as the ratio of the current to the surface area whose plane is normal to the direction of the charge motion. It is denoted by $\vec{J}$ and its unit is ampere metre². Also

$$J = \frac{I}{A}$$

Thus current density can also be defined as the current flowing through a cross-section of unit area.

Ohm's Law

In differential form ohm's law may be written as

$$\vec{J} = \sigma\vec{E}.$$

Thus according to Ohm's law the field strength within a conductor is proportional to the current density. (σ is the conductivity of the material of the conductor.

Continuity equation

The relation $\Delta.\vec{J} + \frac{\partial Qv}{\partial t} = 0$ is known as the law of conservation of charge in differential form. When steady currents are involved

$$\frac{\partial Qv}{\partial t} = 0$$

hence $$\Delta . \vec{J} = 0$$

This equation is called continuity equation for steady current.

Joule's Laws

States that at a point within a conductor which carries direct current, the time rate at which heat is being generated per unit volume is given by

$$\frac{dp}{dv} = \vec{E} . \vec{J}$$

Magnetostatics

Electric charges at rest produce an electric field called electrostatic field. Electric charges in motion, i.e., currents produce magnetic field. The problem of magnetic field of stationary current is called magnetostatics.

Ampere's Law of Force

States that the force $\vec{F}$ between two parallel wires carrying currents I_1 and I_2 is proportional to $\frac{I_1 I_2}{r^2}$ where r is the distance between the centres of the wire.

Biot Savart Law

At any point P the magnitude of the magnetic, field intensity produced by the differential is proportional to the product of current, the magnitude of the differential length and the sine of the angle lying between the filament and a line connecting the filament of the point P where the field is desired. The magnitude of the magnetic field intensity is inversely proportional to the square of the distance from element to the point P.

The direction of the magnetic field intensity is normal to the plain containing the differential element and the line drawn from the filament to the point P. Of the two possible normals, that one is to be chosen which is the direction of progress of a right handed screw turned from dl through the smaller angle to the line from the filament to P.

Using rationalised mks units, the constant of proportionality is $\frac{1}{4\pi}$. Mathematically

$$\vec{B} = \frac{\mu_0 I}{4\pi} \int \left(\frac{dl \times \vec{r}_1}{r^2} \right)$$

Magnetic Flux Density

For an arbitrary surface ds bounded by a closed contour s, total magnetic flux ϕ passing through the surface is given by

$$\phi = \int_s \vec{B}\, ds$$

The unit of ϕ is webers while that of magnetic flux density $\vec{B}$ is weber/metre^2.

Ampere's Circuital Law

According to this law, the line integral of B about closed contour in a region containing steady current, is equal to μ_0 times the total net current flowing through the contour of integration. Mathematically,

$$\int_0 \vec{B}\, d\vec{s} = \mu_0 \int_s \vec{J} d\vec{s}$$

Magnetic Field Intensity

The well-known relation is

$$\vec{B} = \mu \vec{H}$$

where $\mu = \mu_0 (1 + x_m)$ is called the permeability of the material. The unit of permeability is henry per metre in MKS system.

Field of a Surface Conductor

Which acts in the normal direction only is related to the surface charge density by the relation

$$\vec{E}_n = \frac{Q_s}{\varepsilon} \text{V/m}$$

Maxwell's Equation

Fundamental equations of electromagnetic theory are

$$\text{div } \vec{D} = \rho$$

$$\text{Div } \vec{B} = 0$$

$$\text{Curl } \vec{H} = \vec{J} + \frac{\partial \vec{D}}{\partial t}$$

$$\text{Curl E} = -\frac{\partial \vec{B}}{\partial t}$$

Maxwell's equations in differential form are

$$\vec{D} = \varepsilon \vec{E}$$

$$\vec{J} = \sigma \vec{E}$$

$$\vec{B} = \mu \vec{H}.$$

All these equations are called constitutive or subsidiary equations.

Maxwell's in integral form are

$$\int_s \vec{D}\, da = \int_v \rho\, dv = Q$$

$$\int_s \vec{B}\, d\vec{a} = 0$$

$$\oint \vec{H}\, b\vec{d} = -\int_s \left(\vec{J} + \frac{\partial \vec{D}}{\partial t}\right) ds$$

$$\oint \vec{H}\, b\vec{d} = -\frac{\partial}{\partial t} \int_s \vec{B}\, ds$$

Relation between $\vec{E}$ and $\vec{D}$ in a uniform plane wave is

$$\frac{\vec{E}}{\vec{H}} = \sqrt{\frac{\mu_0}{\varepsilon_0}}$$

Poynting Theorem

States that vector product of electric intensities $\vec{E}$ and magnetic field intensity $\vec{H}$ at any point is a measure of the rate of energy flow per unit area at that point. Mathematically

$$\vec{P} = \vec{E} \times \vec{H}$$

The direction of flow is perpendicular to $\vec{E}$ and $\vec{H}$ in the direction of the vector.

Reflection of Plane Wave by a Perfect Conductor

When an electromagnetic wave travelling in one medium strikes upon a second medium the wave in general will be partially transmitted and partially reflected.

(i) When the incidental wave is normal (travelling in x direction), the strength of electric field E(T) x is given by the equation

$$E_T(x, t) = -2E_i \sin \beta_x \sin \omega t$$

(if the medium in which the wave is travelling is assumed to the perfect, with P = Jβ).

Also $E_r(x, t) = — 2H_1 \cos \beta_x \cos \omega t.$

(ii) When the wave is obliquely incident on the interface between two media, two cases may arise.

(a) Horizontal Polarization

In such a case total electric field E_t is given by the equation

$$\dot{E}_t = 2j\, E_i \sin \beta z . z e^{-j\beta \sin\theta}.$$

where $\beta_z = \beta \cos \theta.$

(b) Vertical Polarization

In such a case the total z-component of electric field strength is given by the equation

$$E_z = — 2j\eta \sin \theta\, H_1 \cos \beta z . z e^{-j\beta y} \gamma.$$

and the total y-component of electric field is given by the equation

$$E_y = — 2j\eta \sin \theta\, H_1 \cos \beta z . z e^{-j\beta y} \gamma.$$

Snell's Law

According to Snell's law of optics we have

$$\frac{\sin \theta_1}{\sin \theta_2} = \sqrt{\frac{\varepsilon_1}{\varepsilon_2}}$$

Brewster Angle is the angle for which there is no reflection and the incident wave is vertically polarized. The Brewster angle is given by the equation

$$\text{ten}\, \theta = \sqrt{\frac{\varepsilon_2}{\varepsilon_1}}$$

Total Internal Reflection

According to it, when a wave is incident from a more dense medium into the less dense medium at an angle equal to or exceeding the critical angle, the wave will be totally internally reflected and will also be accompanied by a surface wave in the less dense medium.

OBJECTIVE TYPE QUESTIONS

Q. 1. For an electric medium, which gives the permittivity ?

(A) $k = \frac{k_0}{k_r}$ (B) $k = k_0 k_r$

(C) $k_0 = k.k_r$ (D) $k_r = k_0 k$.

Q. 2. Which is the relation for electric field intensity ?

(A) $E = \frac{Q}{4\pi r}$ (B) $E = \frac{Q}{4\pi r^2}$

(C) $E = \frac{Q}{4\pi \varepsilon r}$ (D) $E = \frac{Q}{4\pi \varepsilon r^2}$

Q. 3. Which of the following equation represent field strength H produced by a conductor carrying current I, at a distance r ?

(A) $H = 2\pi r\, I$ (B) $H = \frac{I}{2\pi r}$

(C) $H = 4\pi r. I$ (D) $H = \frac{I}{2\pi r}$

Q. 4. Which of the following is NOT Maxwell's equation?

(A) $B = \mu H$ (B) $E = \frac{D}{\varepsilon}$

(C) $E = \frac{J}{\sigma}$ (D) $E = \varepsilon D$.

Q. 5. When B is the magnetic flux density B due to wire carrying current I, then according to Biot-Savert's law we have,

(A) $B = \int \frac{I\,dl\,iR}{r}$ (B) $B = \int \frac{I\,dl\,iR}{r^2}$

(C) $B = \mu_0 \oint \frac{I\,dl\,iR}{r^2}$

(D) $B = \frac{\mu_0}{4\pi} \oint \frac{I\,dl\,iR}{q^2}$

Q. 6. A parallel polarized wave is incident from air to paraffin. If ε_r for paraffin is 2, the Brewster angle will be nearly equal to

(A) 45° (B) 40°

(C) 60° (D) 55°.

Q. 7. When n_1 and n_2 are the refractive indices of two mediums, then according to Snell's law

(A) $\frac{n_1}{n_2} = \frac{\sin\theta_r}{\sin\theta_i}$ (B) $\frac{n_1}{n_2} = \frac{\sin\theta_i}{\sin\theta_r}$

(C) $\frac{n_2}{n_1} = \frac{\cos\theta_i}{\cos\theta_r}$ (D) $\frac{n_2}{n_1} = \frac{\cos\theta_r}{\cos\theta_i}$.

Q. 8. Absolute permeability μ is equal to

(A) $4\pi \times 10^{-7}$ H/m (B) $4\pi \times 10^{-12}$ H/m

(C) 8.854×10^{-7} H/m (D) 8.854×10^{-12} H/m.

Q. 9. The absorption of radio waves by the atmosphere depends on which of the following ?

(A) The polarization of atmosphere

(B) The polarization of the waves

(C) Their frequency

(D) Their distance from the transmitter.

Q. 10. Plasma is

(A) A medium dominated by helium

(B) A solid dielectric material

(C) A gaseous medium

(D) A ray of electrons.

Q. 11. Circular polarization is said to take place when

(A) The magnitude of the two waves is the same

(B) The phase of the two waves is the same

(C) The magnitude is the same and the phase difference is zero

(D) The magnitude is the same and the phase difference is 90°.

Q. 12. The velocity of electromagnetic waves in a dielectric medium $\varepsilon_r = 4$ is equal to

(A) 3×10^8 metres/second

(B) 1.5×10^8 metres/second

(C) 6×10^8 metres/second

(D) 12×10^8 metres/second.

Q. 13. A plane wave travelling in air is normally incident on a block of paraffin with εr = 2.2. The reflection coefficient will be equal to

(A) — 0.195 (B) — 0.255

(C) — 0.456 (D) — 0.557.

Q. 14. For different loads the range of the values for VSWR is

(A) 0 to 0.5 (C) 0.5 to 1

(B) 0 to 1 (D) 1 to ∞.

Q. 15. Two dielectric mediums of which medium 1 is free space and medium 2 has $\varepsilon_2 = 4\varepsilon_0$ and $\mu = \mu_0$. The reflection coefficient for oblique incidence $\theta_i = 30°$ for perpendicular polarization will be

(A) — 0.118 (B) — 0.381

(C) — 0.566 (D) — 0.777.

Q. 16. When the distance between two charges is doubled, the force between them will be equal to

(A) Four times (B) Double

(C) Half (D) One fourth.

Q. 17. If the conductivity of copper is 5.8×10^7 mho/m and its relative permeability and permittivity is unity, then its refractive index at 10 MHz be equal to

(A) 2.27 (B) 22.7

(C) 227 (D) 2270.

Q. 18. A sinusoidal plane wave is transmitted through a medium whose breakdown strength is 30 KV/m and whose relative permeability is 4

The mean possible r.m.s. power flow density will be

(A) 2.39 kW/metre (B) 23.9 kW/metre

(C) 239 kW/metre (D) 2390 kW/metre.

Q. 19. Which of the following is ferromagnetic material?

(A) Copper (B) Palladium

(C) Silver (D) Cobalt.

Q. 20. The relative permeability is considered as unity for

(A) Bismuth (B) Vacuum

(C) Air (D) Cobalt.

ANSWERS

1. B,	2. D,	3. D,	4. D,	5. D,	6. D,
7. A,	8. A,	9. C,	10. C,	11. D,	12. B,
13. B,	14. D,	15. B,	16. D,	17. C,	18. D,
19. D,	20. B.				

Chapter 7

Electronic Instrumentation

STUDY MATERIAL

Cathode Ray Oscilloscope

It is an equipment used for checking electronic circuits because it can show the waveform of an applied voltage. The fluorescent screen of the cathode ray tube, displays a graph of voltage amplitude changes with respect to time. The vertical axis represents voltage while the horizontal axis is linear time base for the vertical signal.

With a sine wave input, the oscilloscope shows sine waves on the screen. With an input of square waves or any other waveform, the screen pattern is a picture of the changes. The number of cycles displayed depends on the frequency of the input

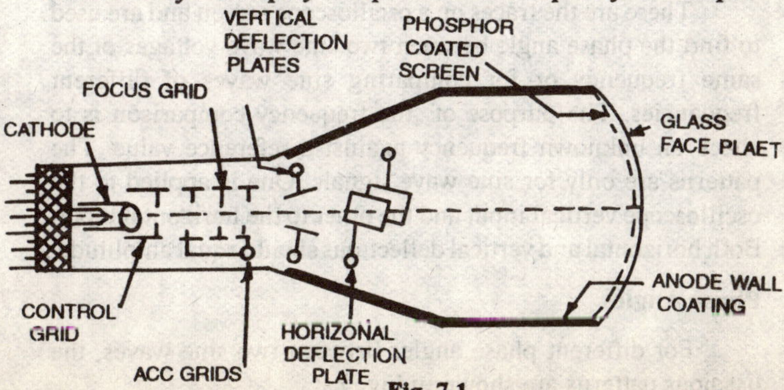

Fig. 7.1.

signal and a horizontal reference frequency. The oscilloscope can measure voltage as also the frequency of the input signal. Oscilloscopes have frequency response from OHz for d.c. upto 100 MHz and above. Thus RF signals can also be measured.

Some Common Screen Phosphors

Phosphor Number	*Colour*	*Persistence*	*Application*
P_1	Yellowish green	Medium	Oscilloscopes
P_4	White	Medium	Monochrome picture tube
P_{22}	Green, blue and red	Medium	Colour picture tubes
P_{31}	Green	Medium short	Oscilloscopes
P_{33}	Orange	Very long	Radar

Lissajous Patterns

These are the traces on a oscilloscope screen and are used to find the phase angle between two sine wave voltages of the same frequency or for comparing sine waves of different frequencies. The purpose of the frequency comparison is to check an unknown frequency against a reference value. The patterns are only for sine wave signals. One is applied to the oscilloscope vertical input and the other to the horizontal input. Both horizontal and vertical deflections should equal amplitude.

Phase Angles

For different phase angles between two sine waves, the lissajous patterns are shown in Fig. 7.2.

For the pattern of an ellipse, the phase angle θ can be found by calculating the ratio of the two lengths B and A shown in Fig. 7.2 at the right end. Thus for phase angle of 45°, the intercept B will be nearly seven tenths of A, so that $\frac{B}{A} = 0.7$ and $\theta = \sin^{-1} 0.7 = 45°$.

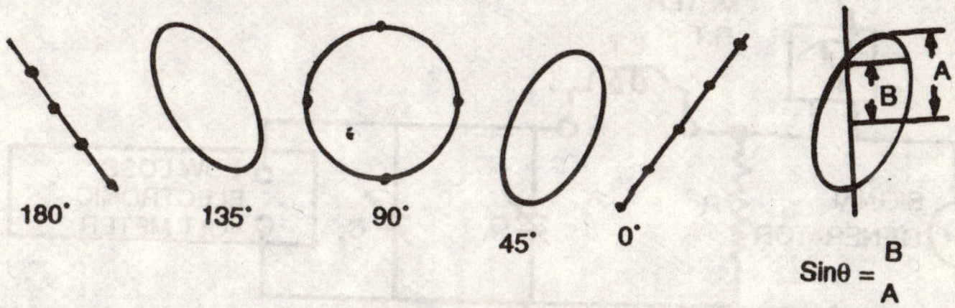

Fig. 7.2.

Frequency Comparison

When vertical and horizontal input signals have specific ratios of frequency, definite patterns are on the screen as shown in Fig. 7.3. Even unknown frequency can be checked by using a frequency generator of known frequency.

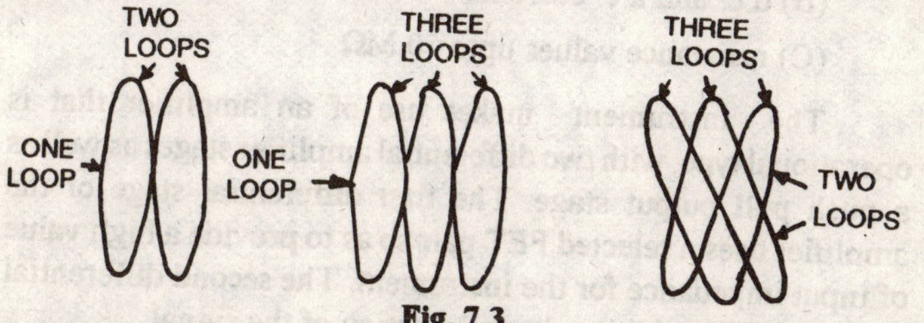

Fig. 7.3.

Q meter. Its principle is shown in Fig. 7.4. In this a RF oscillator is used which generally covers the frequency range from 50 KHz to 70 MHz, in different ranges. From this oscillator, a small voltage (precisely caliberated) is supplied to the circuit comprising the coil under test. The voltage across the internal capacitors is indicated by an electronic voltmeter, of high input impedance, calibrated directly in terms of values of Q.

Fairly good accuracy of the order of $\pm 2.5\%$ to $\pm 5\%$ can be achieved upto 20 MHz.

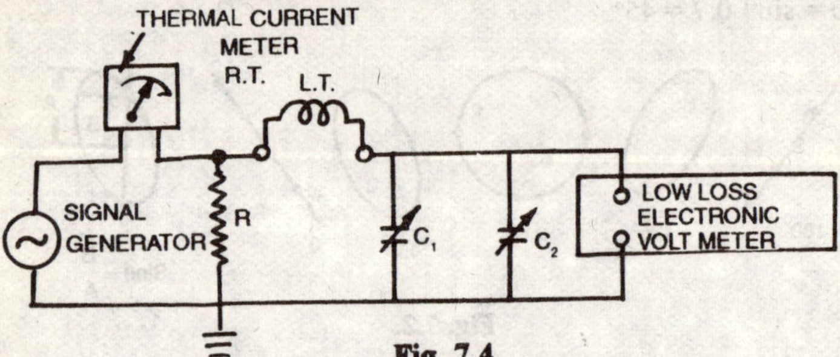

Fig. 7.4.

Electronic Multimeter

This instrument can be used to measure

(A) d.c. and a.c. voltage

(B) d.c. and a.c. currents

(C) resistance values upto 50 MΩ.

The instrument makes use of an amplifier that is operational type, with two differential amplifier stages as well as a push pull output stage. The first differential stage of the amplifier uses a selected FET pair so as to provide a high value of input impedance for the instrument. The second differential stage provides additional amplification of the signal.

The amplifier is preceded by an attenuator for different ranges of voltages, currents and resistances. The instrument can measure d.c. voltages 30 KV and a.c. voltages upto 16V r.m.s. in the frequency range of 100 KHz to 800 MHz.

Vacuum Tube Voltmeter

A block diagram for the meter is shown in Fig. 7.5. In this a cathode follower input circuit is used. This provides high input impedance and is followed by a four stage video amplifier grouped into two pairs with feedback applied individually with each pair. The output from the amplifier is connected to the indicating meter through a full wave rectifier. An additional cathode follower stage is provided to facilitate the instrument to be used as a wide-band amplifier having a frequency response comparable with that of the voltmeter.

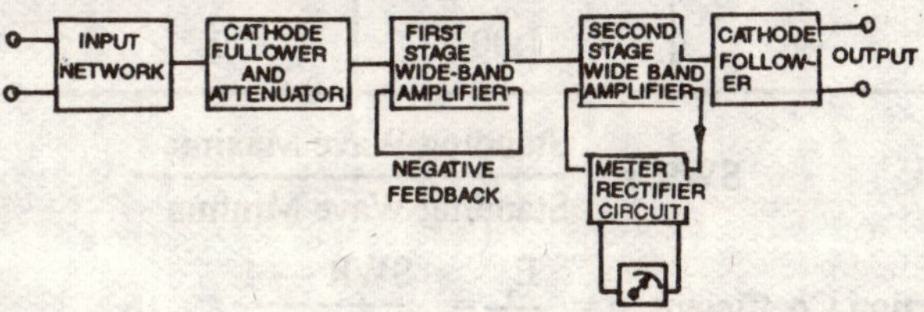

Fig. 7.5.

Microwave Measuring Techniques

For this a complete line of microwave coaxial and waveguide measuring equipment is used. Important parameter that generally needs to be measured is impedance. The impedance of a microwave component is a measure of its input or output match to a given transmission line. The impedance considerations are important to ensure that those components may be further assembled into larger systems without undue effects from mismatch and other reflections. Impedance matching ensures maximum power transfer, low interactions of reflected signals and minimisation of standing waves.

Any mismatch of impedance, on microwave transmission, would cause reflected waves which interact with the incident wave to form a standing wave pattern. Slotted line method is generally used to measure Standing Wave Ratio (SWR). However, reflectometer and directional bridge can also be used.

Relation between S.W.R. | ρ | and Return Loss

SWR	$|\rho|$	*Return loss*
1.00	0.00	∞
1.22	0.10	20 dB
2.00	0.33	9.6 dB
∞	1.00	0

$$\text{SWR} = \frac{\text{Standing Wave Maxima}}{\text{Standing Wave Minima}}$$

$$\text{Reflection Coefficient, } \rho = \frac{E_r}{E_i} = \frac{\text{SWR} - 1}{\text{SWR} + 1}$$

Slotted Line Method

A slotted line may be coaxial or waveguide as needed by the device under test. When loosely coupled to the RF field in the line, the probe, when moved along the line, senses the relative field strength of the standing wave pattern. A SWR indicator is used to measure SWR as shown in Fig. 7.6.

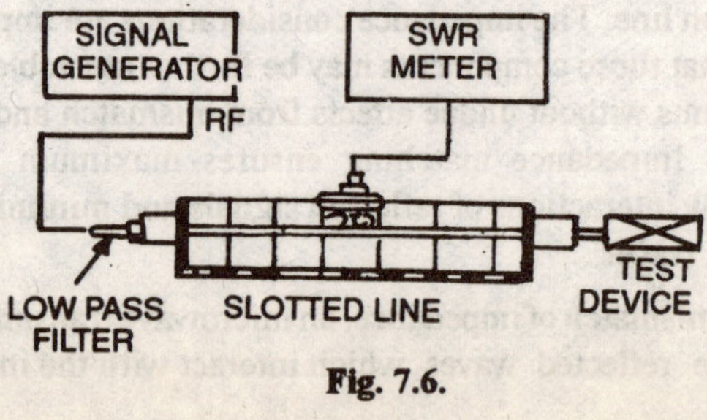

Fig. 7.6.

This method is quite accurate specially at high frequencies because of inherent low residual SWR. This method can also be used to measure phase characteristics by referring the impedance to a shorting plane.

OBJECTIVE TYPE QUESTIONS

Q. 1. The resistance of a circuit is measured by measuring current and power flow through the circuit. If the limiting errors in the measurement of power and current are respectively ± 1.5% and ± 1.0%, the limiting error in the measurement of resistance will be

(A) ± 1.5% (B) ± 2.5%

(C) ± 3.5% (D) ± 5.5%.

Q. 2. The nominal value of a resistor is a 0Ω ± 0.1%. A voltage is applied across the resistor and power consumed is estimated by the relation

$$P = \frac{E^2}{R}$$

The measured values of E and I are

E = 100V ± 1%

I = 10A ± 1%

The uncertainty in the power determination will be nearly

(A) 0.5% (B) 0.23%

(C) 3.2% (D) 23%.

Q. 3. A set of ten independent measurements is given below:1.570, 1.597, 1.591, 1.562, 1.577, 1.580, 1.564, 1.586, 1.550 and 1.575

The arithmatic mean is equal to

(A) 1.501 (B) 1.5156

(C) 1.5704 (D) 1.5752.

Q. 4. Which of the following magnetic material has least value of $(B.H.)_{max}$?

(A) Carbon Steel (C) Cobalt steel

(C) Alnico (D) Alcomax.

Q. 5. Manganin is an alloy of

(A) Copper, manganese and nickel

(B) Copper, zinc and lead

(C) Copper, aluminium and chromium

(D) Copper, chromium and cadmium.

Q. 6. Which is a Major application of constantan ?

(A) Precision resistances (B) Thermocouples

(C) Transistors

(D) Heater elements for valves.

Q. 7. A Ryall crest voltmeter is used to measure which are?

(A) Peak voltage (B) d.c. voltage

(C) All of the above. (D) r.m.s. voltage.

Q. 8. What are Helipots ?

(A) Multi-turn potentiometers

(B) Defects in metallic tranducers

(C) Primary transducers for x-ray measurements

(D) None of the above.

Q. 9. The gauge factor for doped crystals is in the range

(A) 0.5 to 1 (B) 1 to 2

(C) 2 to 20 (D) 100 to 5000.

Q. 10. A nixie tube

(A) is a display device

(B) is a a.c. to d.c. converter

(C) is a d.c. to a.c. converter

(D) is a device for reducing distortion.

Q. 11. LED is

(A) A PN junction which emits light when current passes in the forward direction

(B) A gaseous discharge glow devices

(C) Photo cell device which emits light

(D) None of the above.

Q. 12. A three digit 0—1 V digital voltmeter will have a resolution of

(A) .15 V (B) $\frac{1}{3}$V

(C) 0.1 V (D) 1 mV.

Q. 13. What is the principle of working of thermocouple vacuum gauge?

(A) Variation of pressure of gases with temperature

(B) Variation of thermal conductivity of gases with pressure

(C) Variation of kinetic energy of gases with pressure

(D) None of the above.

Q. 14. An accelerometer has a seismic mass of 0.05 kg and a spring constant of 3×10^3 N/m. Maximum mass displacement is ± 0.02 m before the mass hits the stop. The maximum measurable acceleration will be equal to

(A) 400 m/s^2 (B) 600 m/s^2

(C) 800 m/s^2 (D) 1200 m/s^2.

Q. 15. The dead zone in a certain pyrometer is 0.125 per cent of span. The calibration is 400°C to 1000°C. What temperature change might occur before it is delected ?

(A) 0.075°C (B) 0.75°C

(C) 4.5° D (D) 5°C.

Q. 16. The transducers that convert the input physical phenomenon into an electrical output in the form of pulses are called

(A) Secondary transducers (B) Tuned transducers

(C) Analogue transducers (D) Digital transducers.

Q. 17. A superheterodyne receiver with an IF of 450 KHz is tuned to a signal of 1200 KHz. The image frequency is

(A) 1750 KHz (B) 2650 KHz

(C) 2100 KHz (D) 2870 KHz.

Q. 18. If the carrier of a 100 per cent modulated AM transmitter is suppressed, the percentage power saving will be

(A) 150 (B) 66.6

(C) 190 (D) 100.

Q. 19. A digital voltmeter has a read-out range from 0 to 9999 counts. The resolution of the instrument in volts when the full scale reading is 9.999 V will be equal to

(A) 1 μA (B) 01 mV

(C) 1 mV (D) 10 mV.

Q. 20. In case of strain gauges, the gauge factor 'K' is related to poissons ratio 'μ'by the relation

(A) $\mu = \frac{k+1}{2}$ (B) $\mu = \frac{k-1}{2}$

(C) $K = 1 - \mu$ (D) $K = 1 + 2\mu$

Q. 21. A decade counter needs

(A) 10 flip-flops (B) 4 flip-flops

(C) 5 flip-flops (D) 2 flip-flops.

Q. 22. The duty cycle of a pulse of width 2 micro sec and repetition frequency 4 KHz is

(A) 0.55 (B) 0.055

(C) 0.008 (D) 0.0006.

Q. 23. Four resistors are connected in series. Their values are 28.4 Ω, 4.25 Ω, 56.605 Ω, and 0.75 Ω with an uncertainty of one unit in the last digit in each. The total series resistance (significant figures only) will be equal to

(A) 90 Ω (B) 79 Ω

(C) 78.5 Ω (D) 79.225 Ω.

Q. 24. A low Q factor has

(A) Flat response

(B) Peaked response

(C) Lower losses

(D) Higher losses and flat response.

Q. 25. A RC circuit consists of a capacitor of 1 μF in series with a resistor of 5 KΩ. A d.c. voltage of 50 V is suddenly applied across the circuit. The value of voltage after 5 milli seconds will be

(A) 10.5 V (B) 25.3 V

(C) 31.6 V (D) 44.5 V.

Q. 26. For a.c. circuits a resistor can be approximated as shown in

(A) Figure A (B) Figure B

(C) Figure C (D) Figure D.

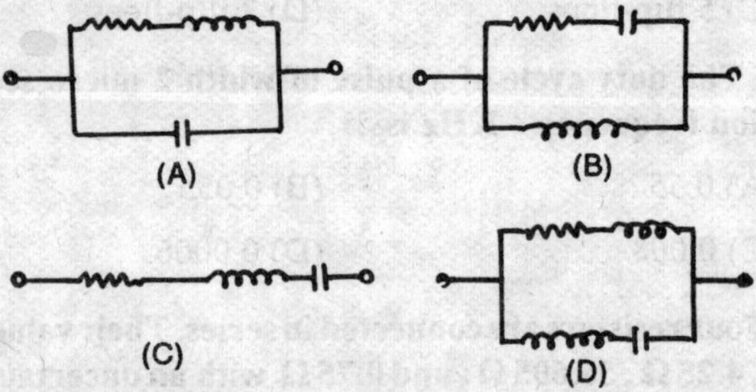

Fig. 7.7.

Q. 27. Fourier equation for the wave shown in Fig. 7.8 can be written as

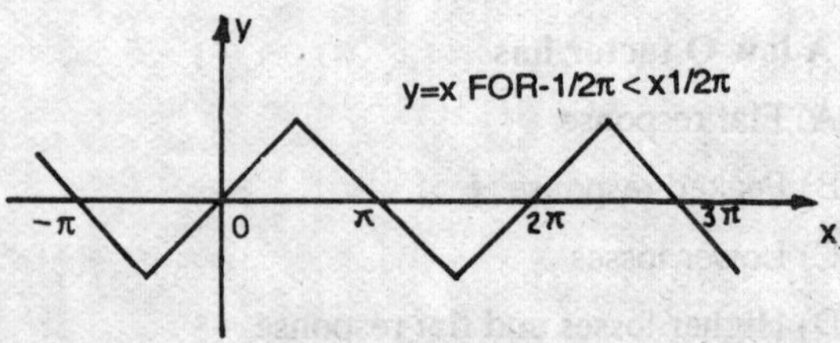

Fig. 7.8.

(A) $y = \frac{4}{\pi}\left(\text{Sin } x - \frac{\sin 3x}{3^2} + \frac{\sin 5x}{5^2} - \ldots\ldots\right)$

(B) $y = \frac{4}{\pi}\left(\cos x - \frac{\cos 3x}{3^2} + \frac{\cos 6x}{5^2} - \ldots\ldots\right)$

(C) $y = \frac{4}{\pi}\left(\cos 2x - \frac{\cos 4x}{4^2} + \frac{\cos 5x}{6^2} - \ldots\ldots\right)$

(D) $y = \frac{4}{\pi}\left(\sin 2x - \frac{\sin 4x}{4^2} + \frac{\sin 6x}{6^2} - \ldots\ldots\right)$

Q. 28. The circuit shown in Fig. 7.9 will be nearly non-reactive when

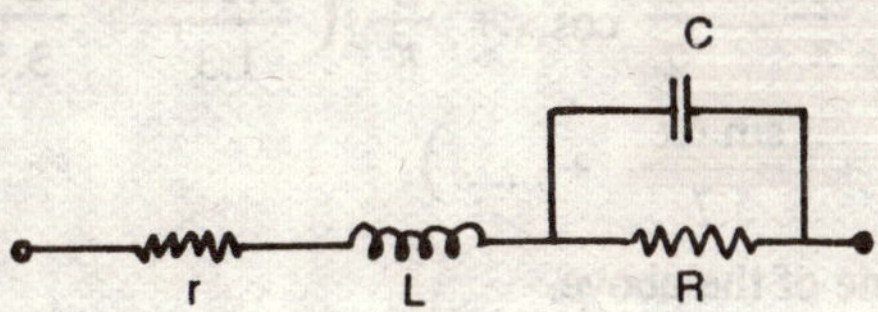

Fig. 7.9.

(A) C = LR (B) $C = LR^2$

(C) $C = L/R^2$ (D) C = L/R.

Q. 29. Fourier equation for the wave shown in Fig. 7.10 can be written as

$y = \sin x$ for $0 < x < \pi$

$y = 0$ for $\pi < x < 2\pi$

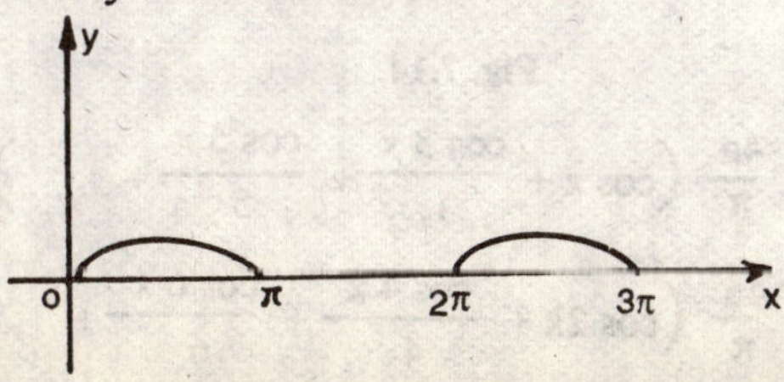

Fig. 7.10.

(A) $y = \frac{1}{\pi} + \frac{1}{2}\sin x - \frac{2}{\pi}\left(\frac{\cos 2x}{1.3} + \frac{\cos 4x}{3.5} + \frac{\cos 6x}{5.7} + \ldots\ldots\right)$

(B) $y = \frac{1}{\pi} + \frac{1}{2}\cos x - \frac{2}{\pi}\left(\frac{\sin 2x}{1.3} + \frac{\sin 4x}{3.5} + \frac{\sin 6x}{5.7} + \ldots\ldots\right)$

(C) $y = \frac{1}{\pi} - \frac{1}{2}\cos x + \frac{2}{\pi}\left(\frac{\cos 2x}{1.3} + \frac{\sin 4x}{3.5} + \frac{\sin 6x}{5.7} + \ldots\ldots\right)$

(D) None of the above.

Q. 30. Fourier equation for the wave shown in Fig. 7.14. can be written as

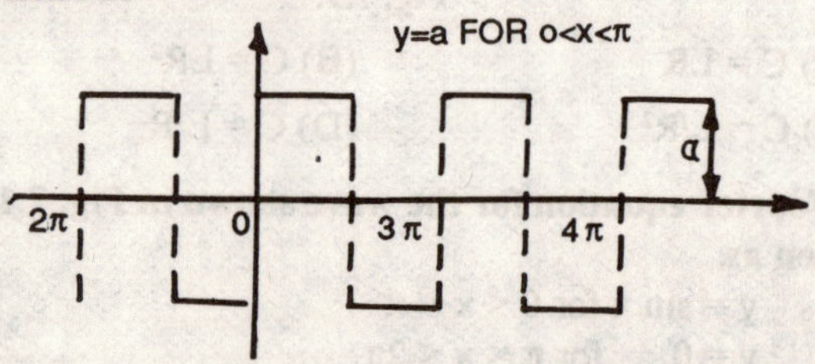

Fig. 7.14.

(A) $y = \frac{4a}{\pi}\left(\cos x + \frac{\cos 3x}{3} + \frac{\cos 5x}{5} + \ldots\ldots\ldots\right)$

(B) $y = \frac{4a}{\pi}\left(\cos 2x + \frac{\cos 4x}{4} + \frac{\cos 6x}{6} + \ldots\ldots\ldots\right)$

(C) $y = \frac{4a}{\pi}\left(\sin x + \frac{\sin 3x}{3} + \frac{\sin 5x}{5} + \ldots\ldots\right)$

(D) $y = \frac{4a}{\pi}\left(\sin 2x + \frac{\sin 4x}{4} + \frac{\sin 6x}{6} + \ldots\ldots\right)$

Q. 31. A µ A 741 C opamp is used in the circuit of Fig. 7.12. The output voltage for the ideal op amp will be

(A) — 5 V (B) — 3 V

(C) — 2.25 V (D) — 2 V.

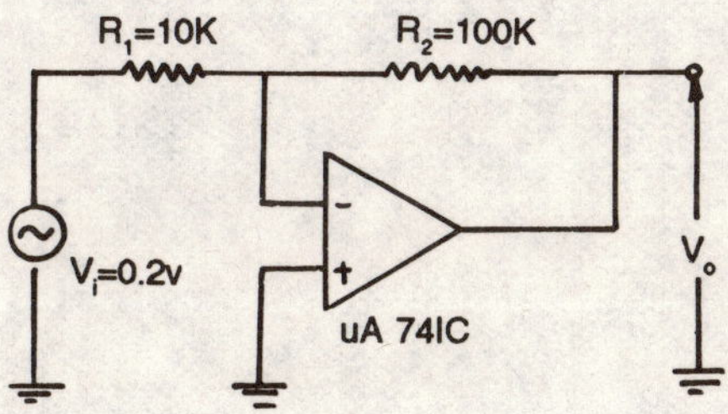

Fig. 7.12.

Q. 32. The display for a sweep frequency of 10 KHz will be

(A) One half cycle (B) One complete cycle

(C) Two cycles (D) Four cycles.

Q. 33. A square wave signal is observed on the oscilloscope to have cycle measured as 6 cm at a scale setting of 30 s/cm. The signal frequency is

(A) 55 Hz (B) 555 Hz

(C) 5550 Hz (D) 55500 Hz.

ANSWERS

1. C,	2. B,	3. D,	4. A,	5. A,	6. B,
7. A,	8. A,	9. D,	10. A,	11. A,	12. D,
13. B,	14. D,	15. B,	16. D,	17. C,	18. B,
19. C,	20. B,	21. B,	22. C.	23. A,	24. D,
25. C,	26. A,	27. A,	28. C,	29. A,	30. C,
31. D,	32. A,	33. C.			

Chapter 8

Materials for Electronic Equipment

STUDY MATERIAL

Resistive Materials

The resistors required for electronic purposes usually have a resistance from 20 Ω to 20 MΩ. Some other desirable characters are given below:

(i) *Stability.* Its resistance value must not change with time.

(ii) *Noise.* The resistor must be free form noise i.e., it must be free from random fluctuations of current when a voltage is applied to it. The irreducible basic noise voltage due to the Johnson effect is always there.

$$\overline{E^2} = 4k\,TR\,\Delta B$$

where $\overline{E^2}$ = mean square of the noise voltage

K = Boltzmann constant

T = ambient temperature °K

R = resistance value in ohms

Δ B = bandwidth in hertz of the detecting system.

Resistors based on conductors-insulator mixtures

A finely divided insulating material can be mixed with a finely divided conducting material to prepare a resistor of required value. The most important of the mixture resistors is the carbon composition, Fig. 8.1. In this carbon black (i.e., carbon in a very finely divided state) is dispersed in resin together with an inert filler like alumina. This resistive mixture is hot pressed into cylindrical shapes with outer insulating sleeves and tinned copper wires are molded in position.

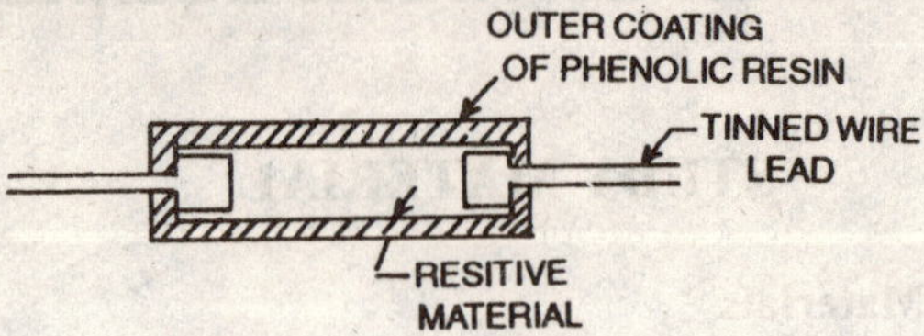

Fig. 8.1.

Tin oxide, a semi-conductor material, doped with antimony is used as the resistive element.

Evaporated Film Resistors

The evaporation of metallic materials is carried out in enclosures at vacuum pressure of about 10^{-8} m of mercury. The subtrate material can be either glass or ceramic. The evaporation of a metallic element is comparatively simple, but most of the familiar resistive materials are alloys, such as Nichrome or constantan. For example, nichrome is an alloy of nickel and chromium and at any temperature they have different vapour pressures. Chromium having higher vapour pressure evaporates first.

Thermistors

It is a component that makes use of the rapid change of the resistivity of a semiconductor with a change of temperature. We have two types of thermistors, one depending on the ambient temperature and the other generating its own temperature. The usual form of a thermistor is made up of a central wire coaxial with an outer sheath of intervening space being packed with the oxide in powder form or in sintered rings with ceramic spacers.

The resistor materials are used as high-power waveguide load materials. High powers may be dissipated by casting the refractory in aluminium with suitably shaped fins to dissipate the heat. The range of power dissipation can be extended still further when the load is surrounded by a jacked through which water is circulated.

Magnetic Materials are used in a wide range of modern electronics equipment. For example, in a digital computer they are used as:

(A) laminations for power transformers and chokes,

(B) transformer laminations for the radio frequency inductors of the read/write recording heads,

(C) the magnetic iron oxide on the magnetic tape which passes adjacently to the read/write recording heads and most important in the computer system,

(D) ferrite elements in the form of small rings with switching properties which form the basis of the central store.

As a class of magnetic materials it is the oxide type, popularly called ferrites, that in view of their insulating nature, have led to some of the most striking advances in the development of the magnetic components.

Most ferrite materials have negative values of anisotropy.

Ferrites for computer applications have rectangular, magnetization curve (see Fig. 8.2.) For a pulse of positive current in a wire threading the core the resultant field will saturate the material at the position A. After the pulse the magnetization will remain saturated at the point B, and further positive pulses will not change the state which remains therefore undisturbed until the negative pulses produces the A' and the B' states. The two magnetized states OB and OB' represent the two clearly

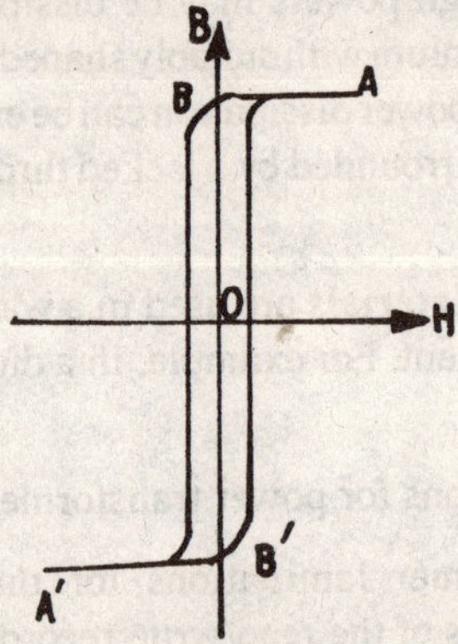

Fig. 8.2.

distinguishable states 0 and 1 necessary for the operation of binary arithmetic.

Because of their insulating properties, ferrites can be used right upto the microwave frequency range where their gyromagnetic properties can be utilised in a number of communications applications. All of these depend on the resonance effect and this in turn is dependent on the nature of the material, its shape, and its physical state.

Although ferrite materials are insulators compared with metallic magnetic materials, care is needed in their preparation so as to develop fully their insulating properties. Being oxides they are quite prone to non-stoichiometry, i.e., they can be produced with small amounts of their oxygen content missing.

OBJECTIVE TYPE QUESTIONS

Q. 1. Select the problem that is associated with superconductors

(A) Maintenance of low temperatures

(B) Materials become highly magnetic

(C) Low strength of materials

(D) Low energy level of electrons does not help in conduction of electric current.

Q. 2. In which is the net magnetic moment is zero ?

(A) Ferromagnetic materials

(B) Ferrimagnetic materials

(C) Antiferrimagnetic materials

(D) None of the above.

Q. 3. Select the heaviest:

(A) Atom (B) Electron

(C) Proton (D) Neutron.

Q. 4. What are ferrites ? They are

(A) Ferromagnetic materials

(B) Ferrimagnetic materials

(C) Ferroelectric materials

(D) Ferri-ferro magnetic materials.

Q. 5. Perminvar is an alloy containing

(A) 45% Cu + 30% Ni + 25% Co

(B) 45% Ni + 33% Co + 25% Cu

(C) 45% Ni + 30% Fe + 25% Co

(D) 45% Co + 30% Ni + 25% Ag.

Q. 6. Human body can sustain electric current only upto

(A) 1 μA (B) 10 μA

(C) 5 mA (D) 35 mA.

Q. 7. Select the one that is visco-elastic

(A) Glass (B) Nylon

(C) Rubber (D) Graphite.

Q. 8. Hastalloy contains

(A) Nickel and copper

(B) Copper and aluminium

(C) Aluminium and nickel

(D) Nickel and molybdenum.

Q. 9. When the temperature of an intrinsic semiconductor is increased

(A) Holes are created in the conduction band

(B) The energy of the atoms increases

(C) The energy of the atoms decreases

(D) The resistance of the semiconductor increases.

Q. 10. At absolute zero temperature all the valence electrons in a semi-conductor

(A) are in the valence band

(B) are in the forbidden gap

(C) are in the conduction band

(D) are free electrons.

Q. 11. A pure semi-conductor material at a temperature of 0°K is

(A) A resistor (B) A conductor

(C) An insulator (D) A source of power.

Q. 12. Charpy test is associated with which property of materials ?

(A) Hardness (B) Malleability

(C) Brittleness (D) Wiredrawing.

Q. 13. Which of the following has highest resistivity ?

(A) Tantalum (B) Manganin

(C) Nichrome (D) Kanthal.

Q. 14. Hall effect is used for all of the following EXCEPT

(A) To find carrier concentration and mobility

(B) To measure conductivity of the material

(C) To detect and measure magnetic fields

(D) To convert P-type semi-conductors into N-type semi-conductor.

Q. 15. The phenomenon of "thermal run away" occurs in

(A) Transistor (B) Magnetic amplifiers

(C) Tube values (D) All of the above.

Q. 16. Materials having large hysteresis loop are considered as suitable for

(A) Electromagnets (C) Commutators

(C) Contactors (D) Permanent magnets.

Q. 17. Which of the following can be expected from a semiconductor ?

(A) Negative temperature coefficient

(B) Almost zero resistance

(C) Infinite resistance

(D) Positive temperature coefficient.

Q. 18. Overhead telephone wires are

(A) ACSR conductors (B) Steel wires

(C) Copper wires (D) Aluminium wires.

Q. 19. The Curie point for Rochelle salt is about

(A) 1000°C (B) 250°C

(C) 24°C (D) Absolute zero.

Q. 20. Which of the following materials has the highest electrical conductivity ?

(A) Steel (B) Silver

(C) Aluminium (D) Zinc.

Q. 21. Piezo electric materials serve as a source of

(A) Microwaves (B) Ultrasonic waves

(C) Musical waves (D) Resonant waves.

Q. 22. A random material should be transparent to

(A) Microwaves (B) IR waves

(C) Both (A) and (B) above (D) None of the above.

Q. 23. Austenite is

(A) F.C.C. iron with nearly 2% (maximum) carbon in solid solution

(B) B.C.C. iron with nearly 0.025% (maximum) carbon in solid solution

(C) B.C.C. iron with nearly 1% carbon in solid solution

(D) None of the above.

Q. 24. Power metallurgy technique is most suitable for making

(A) Large machine base

(B) Transistors

(C) Ferrite magnets

(D) Cutting tools.

Q. 25. Which one of the following is closest to the pure form of iron ?

(A) Mild steel (B) Gray cast iron

(C) Nodular cast iron (D) Wrought iron.

Q. 26. In ACSR conductors

(A) Current flows through steel section only

(B) Aluminium strands takes tension

(C) Steel reinforcement takes major tensile load

(D) Steel takes care of thermal expansion.

Q. 27. "The ratio of thermal conductivity of electrical conductivity of a metal M decreases at low temperatures and the values tends to be zero at zero degrees kelvin." From the above statement, it can be concluded that

(A) Thermal conductivity of metal M is constant

(B) Electrical conductivity of metal varies with temperature only

(C) At absolute zero the thermal conductivity of metal M is zero

(D) At absolute zero, the electrical conductivity of metal M is infinitely large.

Q. 28. Araldite is a (an)

(A) Vegetable oil resin (B) Thermosetting plastic

(C) Epoxy resin (D) A variety of PVC.

Q. 29. The ratio $\frac{\text{resistivity of manganin}}{\text{resistivity of silver}}$ is approximately

(A) 2 (B) 9

(C) 16 (D) 32.

Q. 30. Ligancy is

(A) The number of atoms (or ions) surrounding and touching a central atom

(B) A covalent bond between two atoms

(C) The angle between the two closest directional bonds of an atom

(D) None of the above.

Q. 31. Figure of merit is used to

(A) Compare the efficiency of thermoelectric materials

(B) Measure the extent of doping of intrinsic semi-conductors

(C) Compare the extent of purity of semi-conductor materials

(D) None of the above.

Q. 32. A non-crystalline polymer which can be stretched to more than twice its original length and which contracts quickly on releasing the load, is called

(A) Copolymer (B) Dilatant

(C) Plastic (D) Elastomer.

Q. 33. Secondary emission results

(A) When temperature of metals is raised to a level above the crystallization temperature

(B) When metals are subjected to strong magnetic fields

(C) When light rays fall on the metal surface

(D) When a high velocity beam of electrons strikes a metal surface.

Q. 34. In case of selenium, under the influence of varying light intensity

(A) e.m.f. is generated due to chemical reaction

(B) e.m.f. is generated due to physical reaction

(C) Electrical conductivity changes

(D) The number of electrons liberates varies.

Q. 35. The fourth state of matter is

(A) Space energy

(B) Laser

(C) Plasma

(D) There are three states only.

Q. 36. Which photoconductor is widely used ?

(A) PbS (B) CdS

(C) Ga As (D) As Br.

Q. 37. Wiedemann-Franz law correlates

(A) Electrical and thermal conductivities

(B) Electron arrangement in shells

(C) Temperature and photoelectric emission

(D) All of the above.

Q. 38. A ductile fracture is usually not preceded by

(A) Large amount of non-recoverable energy absorption

(B) Plastic flow

(C) Deformation

(D) Noise.

Q. 39. Noise voltage of resistors due to the Johnson effect is given by

(A) $E^2 = KTR$ (B) $E^2 = 4\ KTR\ \Delta\ B$

(C) $E^2 = 14\ KT^2R\ \Delta\ B$ (D) $E = KTR\ \Delta \cdot B$.

Q. 40. Which of the following dielectric material has the lowest permittivity ?

(A) Titania (B) Ferric oxide

(C) Lead glass (D) Silica.

ANSWERS

1. A,	2. C,	3. A,	4. A,	5. C,	6. D,
7. C,	8. D,	9. B,	10. A,	11. C,	12. C,
13. D,	14. D,	15. B,	16. D,	17. B,	18. B,
19. C,	20. B,	21. B,	22. A,	23. A,	24. D,
25. D,	26. C,	27. D,	28. B,	29. D,	30. A,
31. A,	32. D,	33. D,	34. C,	35. C,	36. B,
37. A,	38. D,	39. B,	40. D.		

Chapter 9

Electronic Communication Technology

STUDY MATERIAL

Electronic Signal

It may be either a.c. or d.c. Any electrical voltage or current that carries some form of intelligence is called **Electronic Signal.**

Vice Signal

It refers to signal that results at the output of a microphone.

In a complex signal there are a number of sinusodial components ranging between 15 Hz to 1500 Hz. It is also known as **Audio range**.

Some other wave forms are square wave signal, triangular signal, sawtooth wave form. All these consist of different sinusodial components having varied frequencies having a definite rotation with one another. e.g. Sawtooth waveform consists of sinusodial components with frequencies which are integral multiple of a fundamental frequency.

Differentiation Circuit with R and C

It is characterised by a very small time constant and the output taken across R. **Small time-constant** here has a special significance. True differentiation occurs only if the time constant R-C is much smaller than the time period of input signal that is to be differentiated.

Integrator

It is a large time-constant circuit with the output taken across C.

Most Commonly Used Signals for Navigational Aids

(i) VOR (Visual Omni Range)

(ii) VAR (Visual Aural Range)

(iii) RDF (Radio Direction Finding)

(iv) ILS (Instrument Landing System)

(v) GCA (Ground Control Approach)

(vi) LORAN (Long Range Navigation)

VOR

In it pilot or captain of the ship gets the exact bearing of plane or ship.

VAR

In it pilot or captain of the ship gets visual as well as aural indication.

ILS

It is exclusively meant for aircraft. In it two types of signals are transmitted from two types of transmitters called (a) Glide path transmitter and (b) Localised transmitter.

In it the position of aircraft in horizontal as also in vertical plane is known.

RDF

The signals that may be coming from any broadcasting system are received by the pilot or the captain of the ship.

LORAN

In it signals transmitted from two different stations—not very far from each other called and slave station.

$$\%\text{age of Modulation in an AM} = \frac{E_m}{E_C} \times 100$$

E_m = Peak amplitude of modulating signal

E_C = Peak amplitude of carrier signal.

$$\text{Power constant of DSB AM signal} = P_C\left(1 + \frac{M^2}{2}\right)$$

$$\text{Where } M = \text{Modulation index}\left(= \frac{E_m}{E_C}\right)$$

Various forms of Amplitude Modulation Systems

These are

(i) SSB (Single Side Band)

(ii) DSB (Double Side Band)

(iii) SSBSC (Signal Side Band with Suppressed Carrier)

(iv) VSB (Vestigial Side Band).

%age of Modulation

It is frequency duration in KHz divided by 75

$$\text{Modulation Index } (M_f) = \frac{\text{Frequency deviation}}{\text{Modulating frequency}}$$

Maximum frequency deviation allowed for FM radio broadcast is 75 KHz.

FM used for sound broadcast of TV signals are allowed a max-deviation of 50 KHz.

FM produced from PM has very stable centre frequency.

FM has great immunity of noise. It is its advantage over AM.

Various Pulse Communication Systems

(i) PAM (Pulse Amplitude Modulation)

(ii) PWM (Pulse Width Modulation)

(iii) PPM (Pulse Position Modulation)

(iv) PCM (Pulse C[illegible]e Modulation).

Of these first three are analog communication systems. PCM is a digital communication system.

Antennas

These are the devices that convert guided e.m. wave into free space e.m. wave and vice-versa.

Polarisation of e.m. Wave

It is the direction of its electrical field component with respect to earth.

Yagi-antena

It is most commonly used antena for TV. It consists of a centre fed half wave dipole, one reflector and one or more directors.

$$\textbf{Standing wave ratio} = \frac{Z_o}{Z_L} \text{ or } \frac{Z_L}{Z_o}$$

$$\textbf{Reflection coefficient} = \frac{Z_L - Z_o}{Z_L + Z_o}$$

Characteristic Impedance of Transmission

$$= \sqrt{\frac{R + J w L}{a + J w C}}$$

where R, L, G and C are series resistances, Inductance, shunt, capacitance respectively.

Maxwells Equations

$$\Delta'' D = \Delta . B = 0$$

$$\Delta \times E = J + \frac{\partial D}{\partial t}$$

$$\Delta \times H = — \frac{\partial D}{\partial t}.$$

Ionosphere

It is the upper part of atmosphere in which some ions are present. This region is formed due to ionization of gases due to solar radiations (e.g. U.V.).

It consists of various layers namely D, E, F_1, F_2.

D is the lowest and F_2 is the highest.

During night F_1 combines with F_2. Most of the communication occurs via F_1, F_2.

Critical Frequency

It is defined as the maximum frequency (for a particular layer) which can get reflected from that particular layer in case of normal incidence.

Maximum Usable Frequency (M.U.F.)

It is defined as the maximum frequency that can be used for communication via a particular layer.

Skip Distance

It is the distance between the transmitter and a point on ground where the e.m. wave reaches first after reflection from ionosphere.

Skip Zone

It is the area which is totally depleted of any reception. Skip Zone is formed when the point on ground, where the e.m. wave after reflection from ionosphere reaches first, lies beyond the maximum communication range of ground-wave propagation.

Optimum Working Frequency (E.W.F.)

It is slightly less than the maximum usable frequency.

Low Level Modulation

The modulation system in which modulating voltage is applied to some preceding stage and not the final RF stage is termed as *Low Level Modulation*. The main advantage of a low-level modulation system is its low audio power requirement.

High Level Modulation

In it modulation occurs in the output power amplifiers with all amplification done in a chain of Class-C amplifiers preceding it. As Class-C amplifiers have a higher efficiency, the overall efficiency of a high-level modulation system is good. The main disadvantage of this system is very high audio power requirement. Amplification of speech signals is done in Class-A amplifiers.

Balanced Modulators

These are used to produce suppressed carrier signals. Armstrong modulator generates FM signals from PM signals.

Various techniques by which output of a transmitter can be changed temporarily or permanently are (i) by reducing H.T. supply (ii) by reducing the coupling between output stage and the feeder (iii) by reducing the no. of stages causing amplification.

Amplitude modulation techniques are sometimes known as A3, A3A, A3H, A3J etc. A3 signifies DSB with full carrier, A3A is SSB with reduced carrier (Pilot-carrier system) A3H is SSB with full carrier, A3J is SSB with suppressed carrier. A3B, is Independent side-band emission (ISB), A5C signifies vestigial Side-band. Lincomplex is an acronym for "linked compressor and expander." It is a system very common with A3B, A3B and A3J. In this system, all frequencies about 2-7 KHz are filtered out to allow for presence of control tone at 2-9 KHz with a band width of 120 Hz.

A Television Camera

It is a device that converts visual information contained in a scene to be televised into corresponding electrical information. It is the most important device at the television transmitter. Various types of television cameras in use are image orthicon, Vidicon, Plumbicon etc. Image orthicon is the representative of all those camera tubes which are based on photoemissive principle where vidicon is the leader of photoconductive camera-tubes. Image-orthicon has large output and thus preferred at the transmitter for commercial broadcast where-as vidicon is preferred for CCTV systems and other such applications where its compactness can be put to use.

A Colour Television Camera

It is nothing but a combination of more than one (2 or 3) monochrome cameras with other accessories. Apart from this, there are a number of other sections to be found at a TV centre.

Television receivers are categorized as 'all-valve', 'hybrid' and 'All solid-state' receivers. The latest 'All-IC-receivers' are an extension of all solid-state receivers only. Local oscillator frequency is greater than the received picture and sound carrier frequency by an amount equal to their respective IF_s. As an example, for channel 4, local oscillator frequency is 101.15 MHz. Also, if a particular channel is spread from AMHz to B MHz (B-A = $7MH_z$ in CCIR standards), then picture carrier = (A+1.25) MHz and sound carrier = (B — 0.25) MHz.

Balun

It is a balancing transformer used for matching either the output impedance of antenna (receiving) with characteristic impedance of transmission line or characteristic impedance of transmission line with receiver's input impedance.

Brightness-control in a TV receiver is located in the grid-cathode circuit of the TV receiver. Brightness varies as the bias is changed. Contrast is located in the output of video amplifier. Contrast variation is achieved by varying the strength of video signal being fed to the cathode-grid circuit of picture-tube. Volume is located in the audio section. All the three are variable resistances. Fine-running control located in the local-oscillator section is a variable capacitance. Also top-linearity control is a present resistor whereas horizontal linearity is generally achieved by a present coil known as HOR COIL. In picture-tube, electro-magnetic deflection is used whereas electrostatic deflection technique is preferred for camera-tubes. Picture-tube anode voltage required varies with screen size. Roughly, it X KV where X is the diagonal dimension of the tube screen in inches. The circuit that generates such a high value of voltage is know as EHT unit.

Radar (Radio Detection and Ranging)

It is used for detecting presence and location of distant objects. It is based on the principle of sending e.m. waves by a scanning antenna and then studying the information content of received echo.

Doppler Effect

According to it, the frequency to transmitted electromagnetic waves is not necessarily the same as the frequency of the same wave when received. If transmitter and receiver are moving towards each other, receiving frequency is larger. This shift in frequency is known as Doppler-shift.

MTI (Moving Target Indicator)

It helps in detection of moving targets. Doppler shift is $\frac{2Vfn}{C}$.

Radar Range

It is given by $(R_{max.})^4$

$$= \left[\frac{P_t G_r A_t \sigma}{(4\pi)^2 (S_{min})} \right]$$

Minimum detectable signal for a typical radar receiver is given by $S_{min} = KT_0 BF_n$ (S/N).

Devices used for Measuring Microwave Power

These are Calorimeter Wattmeter, Bolometer, Thermocouples etc. Each type of power measuring device is suited for some specific applications. Another device that is used for measuring voltage-standing wave ratio is called *Travelling Detector*.

A geostationary satellite orbits the earth in 24 hr. i.e. it appears to be stationary to an observer on earth. All communication satellites are geostationary.

DOMSAT (Domestic Satellites)

These are low altitude satellites meant for providing national coverage for all sorts of communications. Velocity (V) of a geostationary satellite in terms of radius of earth (R) and height of the satellite above earth's surface (h) is given by $V=R\sqrt{9/R+h}$ where (g) is acceleration due to gravity. Geostationary satellites generally orbit in elliptical equatorial orbit whereas low altitude satellites orbit in elliptical polar orbits.

Actuating Current

It is defined as the minimum current at which armature is attracted whereas releasing current is the maximum current for which armature gets released.

Various systems in which d.c. telegraph circuits operate are (i) Simplex (ii) Duplex (iii) Quadruplex (iv) Diplex. In simplex-system, only one message is sent over the line at a time. In quadruplex system, two messages in one direction and two in the other are sent simultaneously over the same line and Differential-duplex system is based on the principle of current balance whereas bridge-duplex system is based on the principle of potential balance.

A scheme of communication where several channels are interleaved and then transmitted together is known as **Frequency Division Multiplex (FDM)** whereas **Time-Division-Multiplex (TDM)** interleaves pulses belonging to different transmissions.

Telephone traffic is measured in **Erlangs**.

BIT

It is a unit information in Binary system while DIT is a unit of information in Decimal system. 1 DIT equal 3.32 bits.

Channel capacity (C) in terms of channel band-width (B) and no. of coding levels (N) under noise free conditions is given by $C=2B \log_2 N$. Under noisy condition, it is given by

$$C = B \log_2 (1 + S/N)$$

Where $\frac{S}{N} \rightarrow$ signal to noise ratio.

OBJECTIVE TYPE QUESTION

Q. 1. Voice signal frequency lies between

(A) $0 \rightarrow 20$ KHz (B) $0 \rightarrow 1$ MHz

(C) 15 Kz $\rightarrow$ 15 KHz (D) None of these.

Q. 2. A complex waveform made up frequency components 1 Hz, 3 Hz, 5 Hz, 7 Hz and 9 Hz. Its fundamental frequency is

(A) 9 Hz (B) 12.5 Hz

(C) 1 Hz (D) Indeterminate.

Q. 3. A sawtooth waveform is made up of

(A) All odd harmonics

(B) All even harmonics

(C) All odd and even harmonics

(D) None of these.

Q. 4. "A square waveform applied to a differentiator circuit gives positive and negative spikes." Is it true ?

(A) Yes (B) No

(C) Not necessarily (D) None of these.

Q. 5. See the circuit shown in Fig. 9.1 below

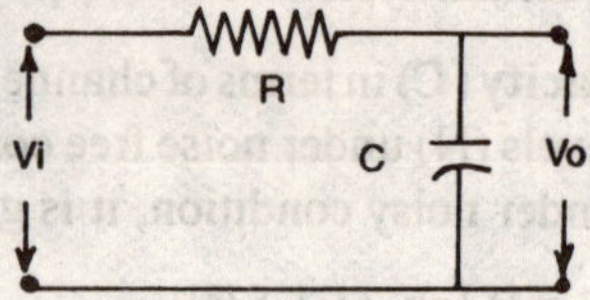

Fig. 9.1.

(A) It is a differentiator (B) It is an integrator

(C) It is an alternator (D) None of these.

Q. 6. An INTEGRATOR circuit should have

(A) Very large time constant

(B) Very small time constant

(C) A time constant that is much smaller as compared to the time occupied by one cycle of applied input

(D) None of these.

Q. 7. Standard LORRN operates in which frequency range

(A) 1.8 to 2.0 MHz (B) 88 to 112 MHz

(C) 108 to 118 MHz (D) None of these.

Q. 8. In all PULSE COMMUNICATION SYSTEMS, carrier

(A) is necessarily a high repetition rate train of pulses

(B) is necessarily a high frequency continuous a.c. signal

(C) is either a train of pulses or a continuous a.c. wave

(D) None of these.

Q. 9. In amplitude modulation

(A) The time gap between adjacent maximum and minimum values of the modulate envelop is not dependent on frequency of intelligence.

(B) Maximum value of intelligence has no effect the time separation of modulated envelope's maximum and minimum.

(C) Separation of successive maximum and minimum of modulated envelope on time-axis is inversely proportional to the frequency of modulating signal

(D) None of these.

Q. 10. In DSB system with carrier when the intelligence varies between 100 Hz to 5 KHz, bandwidth is

(A) 10 KHz (B) 9.8 KHz

(C) 200 Hz (D) Indeterminate.

Q. 11. Signal at the output of an AM modulator is given by e = 5.3 (1+0.64 sin 6280t) sin 10^6t. Depth of modulation for it is

(A) 0.64% (B) 0.80%

(C) 64% (D) 80%

Q. 12. In FREQUENCY MODULATION

(A) Frequency of the carrier remains unchanged

(B) Carrier frequency changes in accordance with the modulating signal amplitude

(C) Carrier frequency changes in accordance with the modulating signal frequency

(D) None of these.

Q. 13. Modulation index in FM is given by

(A) $m_f = \dfrac{75}{\text{Freq. deviation in KHz}}$

(B) 75 KHz is the modulation index

(C) $\dfrac{\text{Freq. deviation in KHz}}{75}$

(D) None of these.

Q. 14. Modulation index (m_f) in FM is

(A) Directly proportional to frequency deviation and inversely proportional to the modulating frequency

(B) Given by $m_f = \dfrac{\text{Frequency Deviation in KHz}}{75} \times 100$

(C) Directly proportional to modulating signal frequency and inversely proportional to frequency deviation

(D) Given by $m_f = \dfrac{\text{Frequency deviation}}{\text{Modulating signal amplitude}}$

Q. 15. Polarization of an electromagnetic wave is

(A) The direction of its magnetic field component with respect to ground

(B) The direction of its electrical field component with respect to ground

(C) The orientation of antenna that generates it.

Q. 16. Is the following figure a valid one ?

(A) It is (B) It is not

(C) It is valid under certain circumstances

(D) None of these.

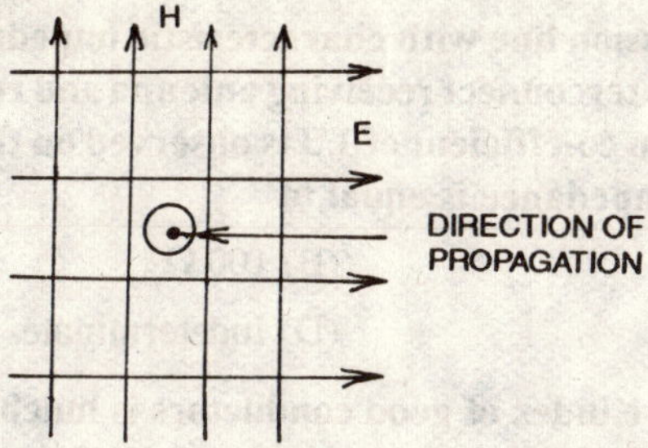

Fig. 9.2.

Q. 17. Is the following figure a valid one ?

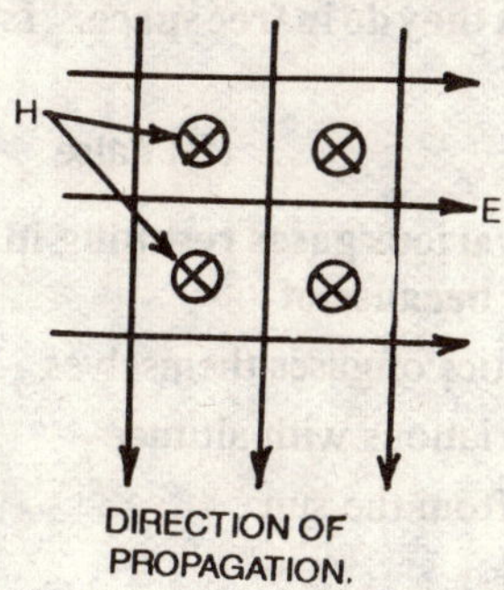

Fig. 9.3.

(A) Yes (B) No

(C) Not necessarily (D) None of these.

Q. 18. Which is most commonly used transmission-line with television system?

(A) Twin-lead

(B) Open-wire with ceramic supports

(C) Coaxial cable

(D) None of these.

Q. 19. A transmission line with characteristic impedance of 300 Ω is used to interconnect receiving antenna and receiver input. A reflection co-efficient of 1/3 is observed on the line. Receiver input impedance is equal to

(A) 150 Ω (B) 100 Ω

(C) 900 Ω (D) Indeterminate.

Q. 20. "Refractive index of good conductors is much larger than that of air" Is this statement true or false ?

(A) True (B) False.

Q. 21. "The electromagnetic waves travels faster in an ionized gas layer than they do in free space." Is this statement true or false ?

(A) True (B) False.

Q. 22. Ionization of various gases resulting in the formation of IONOSPHERE is because of

(A) Characteristics of gases themselves

(B) Pressure variations with altitude

(C) Radiations from the sun

(D) None of these.

Q. 23. A modulated signal (AM) is having depth of modulation equal to 80%, then feeder current under modulated conditions is greater than that under unmodulated conditions by

(A) 4% (B) 10%

(C) 15% (D) None of these.

Q. 24. Choose the wrong statement:

(A) One RF section has only two tuned circuits

(B) One IF section has only four tuned circuits

(C) One IF stage has more selectivity than one RF stage

(D) Selectivity provided by one IF stage is approximately four times that of one RF stage.

Q. 25. Fidelity, in a communication receiver, is provided by

(A) Mixer stage

(B) Detector stage

(C) Various amplifier sections

(D) Audio stage.

Q. 26. Choose the correct statement:

(A) Higher selectivity means larger bandwidth

(B) Higher selectivity means larger gain

(C) A loudspeaker may be rated as 2.5"

(D) Higher sensitivity implies larger gain.

Q. 27. In a ratio-detector

(A) The linearity is worse than in a phase discriminator

(B) Stabilizations against signal strength variations is provided

(C) The output is twice that obtainable from a similar phase discriminator

(D) The circuit is the same as in a discriminator except that diodes are reversed.

Q. 28. Television cameras A and B have tubes which are 3" and 4.5" long respectively. When a lens of known focal length is used with the tubes, then

(A) Angle-of-view of A is greater than that of B

(B) Angle-of-view of B is greater than that of B

(C) Both of them have an identical angle of view

(D) None of these.

Q. 29. Video signals have a frequency range that is spread over approximately

(A) 9 octaves (B) 18 octaves

(C) 4 octaves (D) None of these.

Q. 30. Where is brightness control located in a TV ?

(A) The audio section

(B) The video section

(C) The grid-cathode circuit of picture-tube

(D) None of these.

Q. 31. Where is fine tuning control present in Television receiver ?

(A) A preset inductance (B) A potentiometer

(C) A variable condenser (D) None of these.

Q. 32. Horizontal oscillator in a TV receiver generates a sawtooth signal of

(A) 15,750 Kz in FCC standards

(B) 15,750 Hz CCIR standards

(C) 15,625 Hz FCC standards

(D) 50 Hz in CCIR and 60 Hz in FCC standards.

Q. 33. Television screen width is always greater than the screen height or in other words rastar width more than rastar height because

(A) Horizontal deflection frequency is higher

(B) More detail is in the horizontal direction

(c) Picture motion is in the horizontal direction

(d) None of these.

Q. 34. A TM wave is characterized by

(A) The absence of magnetic component in the direction of propagation

(B) The electric field component wholly transverse

(C) The absence of electric field component in the direction of propagation

(D) None of these.

Q. 35. "A TEM wave cannot exist within a hollow metallic pipe." Is it true ?

(A) Yes (B) No

(C) Not necessarily (D) None of these

Q. 36. Transmission of a wave in a closed waveguide is possible only when

(A) The frequency of operation is less than a certain critical value

(B) The phase-shift (β) is real positive

(C) The phase-shift (β) is zero

(D) None of these.

Q. 37. Choose the correct expression

(A) $A_c = \dfrac{2}{\sqrt{\left(\frac{m}{A}\right)^2 + \left(\frac{m}{A}\right)^2}}$

(B) $\lambda_c = \dfrac{2}{\sqrt{\left(\frac{m\pi}{A}\right)^2 + \left(\frac{n\pi}{A}\right)^2}}$

(C) $\lambda_c = \dfrac{u}{f_c}$ (D) None of these.

ANSWERS

1. C,	2. C,	3. C,	4. A,	5. A,	6. A,
7. D,	8. C,	9. A,	10. C,	11. B,	12. C,
13. A,	14. A,	15. B,	16. A,	17. B,	18. C,
19. A,	20. A,	21. A,	22. B,	23. C,	24. D,
25. D,	26. D,	27. A,	28. D,	29. C,	30. A,
31. B,	32. D,	33. A,	34. A,	35. C,	36. A,
37. C.					

Chapter 10

Electronic Circuits

STUDY MATERIAL

Electronic Amplifier

It is a circuit arrangement that gives amplification of applied A.C. signals.

Direct Coupled Differential Amplifier

These amplifiers are characterised by a very high differential gain (ideally infinite), very high bandwidth, very high input very low input resistances. These amplifiers when slightly modified are called operational amplifiers.

Opamps

These are extensively used for analog computation, performing the functions of addition, substraction, multiplication, division, differentiation, integration, etc.

Important Parameters of Opamps

Common-Mode-Rejection-Ratio (CMRR), Power Supply Rejection Ratio (PSRR), Common Mode Rejection (CMR), input and output offset currents, input bias current, input and output offset voltages etc. are some of the most important performance parameters of an opamp, other parameters are its differential gain, common mode gain, bandwidth and input, output impedances.

Common-Mode-Rejection-Ratio (CMRR)

It is defined as ratio of differential gain (A_d) to common-mode gain (A_c).

Differential Gain (A_d)

It is the ratio of single ended output to differential input.

Common Mode Gain

It is the ratio of single ended output to common-mode input.

Common-Mode-Rejection (C.M.R.)

It is (CMRR) expressed in decibels.

Various Modes in which transistor or valve amplifiers are operated

The following three configurations are commonly used viz., Common Emitter (CE), Common Base (CB), Common Collector (CC) in case of transistor amplifiers and common-cathode, common grid and common plate in case of valve amplifiers.

Common Base (Common Grid)

It is characterised by highest voltage gain, lowest current gain, highest output impedance and lowest input impedance. This type of configuration is extensively used as a constant current source, as an impedance match between a low impedance source and high impedance lead. Current gain of CB configuration is also termed as (α) alpha. Its value is slightly less than unity.

Common Collector Configuration or an Emitter Follower

It is characterized by highest current gain and lowest voltage gain, lowest output impedance and highest input impedance. Voltage gain of an emitter follower is slightly less than unity. It is extensively used as a "buffer stage" to match high impedance source to a low impedance load without altering circuit conditions.

Common Emitter

It is the most versatile of all configurations. It is the only configuration that is capable of providing both voltage and current gains greater than unity and thus maximum power gain. Its input and output impedances are neither too small nor too large and are almost identical. This facilitates the cascading process. At present, almost the entire electronic amplifier circuitry part is based on this configuration.

Push-pull type of Amplifier

In this type all even order harmonics are cancelled. Since second order harmonic distortion is the most dominant one, a push-pull amplifier virtually gets rid of entire harmonic distortion. Because of this that most of audio power amplifiers are push-pull type.

Class-A Amplifier

In it current flows through the active device for whole of the input excitation.

Class-B

In it current flows for half of the input excitation.

Class-C

In it flows for a period less than even one half of the input signal cycle.

Oscillator

It is nothing but a device that generates a.c. signals without any input a.c. excitation. The energy present at the output is derived from d.c. source used for active devices.

Barkhausen Criterion for Oscillators

According to it, loop gain must at least be unity for oscillations to occur and loop phase-shift must be zero. The latter condition determines the frequency of oscillations R-C oscillators are preferred at audio frequencies, L-C oscillators are preferred at radio frequencies whereas crystal oscillators are used for having extremely stable output frequency. Most commonly used L-C oscillators are the Hartley oscillator, the Colpitt oscillator, the Clapp's oscillator, tuned plate tuned grid oscillator. Some of the popular audio oscillators are R-C phase shift oscillator, Weinbridge oscillator etc.

Components of Typical Power Supply

It comprises of a transformer, rectifier, filter and a regulator.

Transformer utilization factor

It is the ratio of d.c. power delivered to load to the a.c. rating of transformer secondary.

Peak Inverse Voltage of a Half Wave Rectifier

It is V_M and $2V_M$ with a capacitor filter, it is V_M for a full wave bridge and $2V_M$ for a conventional full wave rectifier. Inductor filters are suitable at high loads while capacitor filters are practical at low loads. Their combination works alright over a wide range. Regulation of a power supply is

$\left[\dfrac{V_n - V_{f.L}}{V_{f.L}} \times 100 \right]$ per cent. Higher the percentage of regulation, poorer the power supply.

OBJECTIVE TYPE QUESTIONS

Q. 1. Choose the correct statement:

(A) $dB = 20 \log \frac{V_0}{V_i}$ (B) $dB = 20 \log \frac{I_0}{I_i}$

(C) $dB = 10 \log \frac{P_0}{P_i}$ (D) $dB = 20 \log \frac{P_0}{P_i}$

Q. 2. Response curve of an amplifier is a graph that shows

(A) Variation of output voltage with input voltage

(B) Variation of output current with input current

(C) Variation of output voltage with frequency

(D) Variation of output current with frequency.

Q.3. An amplifier has a power gain of 3 dB. This signifies that

(A) $P_0 = 4 P_i$ (B) $P_0 = 3P_i$

(C) $\frac{V_0}{V_i} = 2 \frac{I_i}{I_0}$ (D) $\frac{2V_0}{V_i} = \frac{I_i}{I_0}$.

Q. 4. In FET amplifiers, input excitation is

(A) A current signal (B) A voltage signal

(C) Either a current or voltage signal

(D) None of these.

Q. 5. Decibel is defined in terms of

(A) A power ratio (B) A voltage level

(C) A current level (D) None of these.

Q. 6. "Power gain of a CB configuration is almost equal to its voltage gain." Is it true ?

(A) Yes (B) No.

Q. 7. A source of 500 KΩ output impedance was to be connected to a load of 10 Ω. The matching device or circuit could be of two types. One was by using a CC (common collector) stage in between and other to have a passive circuit having input and output impedances of 500 KΩ and 10 Ω respectively. Finally, CC stage was preferred even though it demanded much more component count and circuit complexity. Why?

(A) Because, it does not introduce any noise

(B) Because, its impedances are extremely stable and do not vary with various parameters

(C) It does not alter circuit conditions because of its voltage gain being unity

(D) None of these.

Q. 8. Mark the reason for (CB) configuration having very small input impedance

(A) Because of its inherent characteristics

(B) That it has a very large resistance gain $\left(= \frac{R_o}{R_i}\right)$

(C) That input is applied between emitter and base and emitter junction is always forward biased for operation

(D) That it is always fed from a low impedance source.

Q. 9. The power handling capability of a given transistor, is dependent upon which of the following ?

(A) Maximum junction temperature and expected ambient temperature

(B) The maximum temperature at which the junction of the device is permitted to operate

(C) The actual ambient temperature expected during operation

(D) None of these.

Q. 10. It is desired to design an amplifier with a voltage gain of 10,000. Output should be inphase with the input. Available CB stage has a gain of 100 per stage and CE has a per stage gain of 25. The designed amplifier shall have

(A) Two cascaded stages of CB configuration

(B) Four cascaded stages of CE configuration

(C) Two cascaded stages of CE configuration

(D) Three cascaded stages of CF configuration followed by an emitter follower stage.

Q. 11. h-parameters of a transistor

(A) Are constant

(B) Vary with temperature

(C) Are dependent upon collector current

(D) None of these.

Q. 12. High frequency response of a transformer coupled amplifier is generally limited by

(A) Transformer's leakage inductance and distributed capacitance

(B) Transformer's primary inductance

(C) Transformer's leakage inductance only

(D) Transformer's winding capacitance only.

Q. 13. Low frequency response of a transformer coupled amplifier is limited by

(A) Transformer's leakage inductance

(B) Transformer's primary inductance

(C) Resistances associated with transformer windings

(D) Transformer interwinding capacitances.

Q. 14. In a voltage series feedback

(A) Output resistance increases while input resistance decreases

(B) Output and input resistances are reduced

(C) Output and input resistances are increased

(D) Output resistance decreases while input resistance increases.

Q. 15. The circuit shown in Fig. 10.1 represents a/an

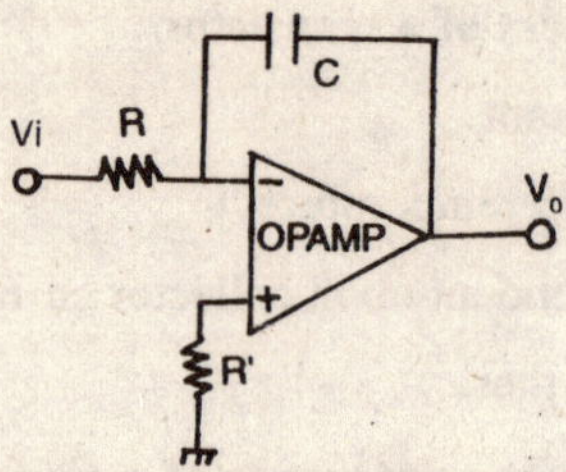

Fig. 10.1.

(A) Integrator (B) Differentiator

(C) High-pass filter (D) None of these.

Q. 16. The circuit shown in Fig. 10.2 represent

(A) A differentiator (B) An integrator

(C) A low-pass filter (D) None of these.

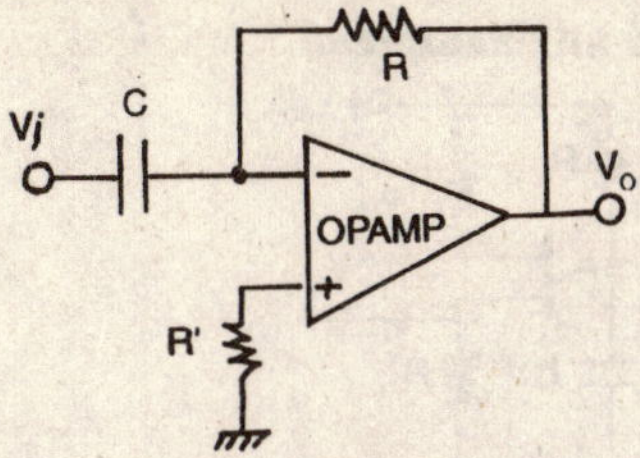

Fig. 10.2.

Q. 17. Out (V_0) in circuit shown in Fig. 10.3 is

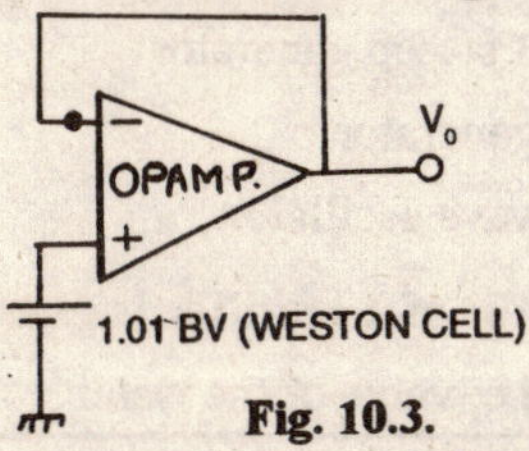

Fig. 10.3.

(A) 1.018 V (B) 0 V

(C) —1.018 V (D) None of these.

Q. 18. Circuit shown in Fig. 10.4 is used as

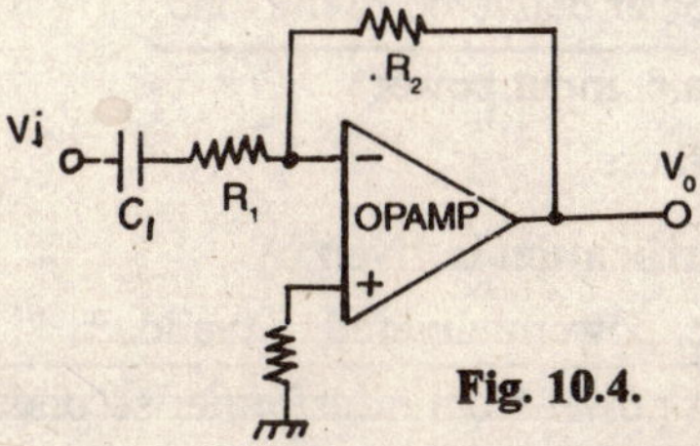

Fig. 10.4.

(A) Inverting amplifier for a.c. applications only

(B) Inverting amplifier for d.c. application only

(C) Inverting amplifier for both d.c. and a.c. applications

(D) None of these.

Q. 19. Fig. 10.5 is an example of

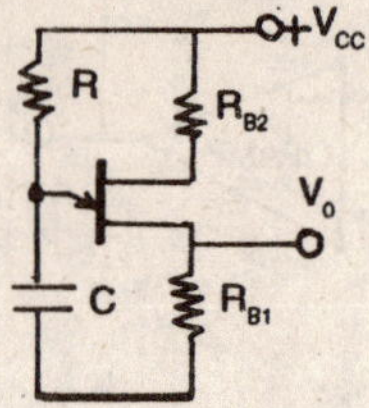

Fig. 10.5.

(A) A voltage time base generator

(B) A current sweep generator

(C) A spike generator

(D) Square-wave oscillator.

Q. 20. Ripple factor of a rectifier is given by

(A) $\dfrac{\text{d.c. value of the water}}{\text{r.m.s. value of a.c. components}}$

(B) $\dfrac{\text{r.m.s. value of a.c. components}}{\text{d.c. value of the wave}}$

(C) $\dfrac{\text{d.c. power delivered to the load}}{\text{a.c. input power}}$

(D) None of these.

Q. 21. Ratio of rectification is given by

(A) $\dfrac{\text{d.c. power delivered to the load}}{\text{a.c. input power from transformer secondary}}$

(B) $\dfrac{\text{a.c. input power from transformer secondary}}{\text{d.c. power delivered to load}}$

(C) $\dfrac{\text{r.m.s. value of a.c. components}}{\text{d.c. value of the wave}}$

(D) None of these.

Q. 22. Percentage voltage regulation may be given by

(A) $\dfrac{V_{no\ load} - V_{full\ load}}{V_{full\ load}} \times 100$

(B) $\dfrac{V_{no\ load} - V_{full\ load}}{V_{no\ load}} \times 100$

(C) $\dfrac{V_{full\ load} - V_{no\ load}}{V_{no\ load}} \times 100.$

Q. 23. Transformer utilization factor of a d.c. power supply may be given by

(A) $\dfrac{\text{Transformer secondary power}}{\text{Transformer primary power}}$

(B) $\dfrac{\text{d.c. power delivered to load}}{\text{a.c. rating of transformer secondary}}$

(C) $\dfrac{\text{d.c. power delivered to load}}{\text{a.c. power from transformer secondary}}$

(D) None of these.

Q. 24. Internal resistance of a power supply can be calculated with the help of

(A) $R_{int} = \dfrac{V_{full\ load}}{I_L}$

(B) $R_{int} = \dfrac{V_{no\ load}}{I_L}$

(C) $R_{int} = \dfrac{V_{no\ load} - V_{full\ load}}{I_L}$

(D) None of these.

Q. 25. An ideal power supply is characterized by

(A) Very large output resistance

(B) Very small output resistance

(C) Zero internal resistance

(D) Infinite internal resistance.

ANSWERS

1. C,	2. C,	3. C,	4. B,	5. A,	6. A,
7. C,	8. C,	9. A,	10. B,	11. C,	12. A,
13. B,	14. B,	15. A,	16. A,	17. A,	18. A,
19. C,	20. B,	21. A,	22. A,	23. B,	24. C,
25. C.					

Chapter 11

Feedback and Control Systems

STUDY MATERIAL

System

It is an arrangement, set or collection of things connected or related in such a manner as to form an 'entirety or whole.

Control

It is usually taken to mean regulate, direct or command.

Control System

It is an arrangement of physical components connected or related in such a manner as to command, direct or regulate itself or another system.

Basic Types of Control Systems

1. Man-made control systems
2. Natural, including biological control systems
3. Control systems whose components are both man-made and natural.

Classification

(i) An *Open-loop* control systems is one wherein the control action is independent of the output.

(ii) A *Closed-loop* control system is one wherein the control action is somehow dependent on the output.

Feed Back

It is that characteristic of closed-loop control systems which distinguishes them from open-loops systems. Feed back allows the output to be compared with the input to the system so that the appropriate control action can be formed as some function of the output and input.

Characteristics of Feed Back

(i) Increased accuracy

(ii) Reduced sensitivity of the ratio of output to input to variations in system characteristics

(iii) Reduced effects of non-linearities and distortion

(iv) Increased bandwidth

(v) Tendency to wards oscillation or instability.

Servomechanism

It is a power-amplifying feedback control system wherein the controlled variable C is mechanical position, or a time derivative of position like velocity or acceleration.

Regulator or Regulating System

It is a feedback control system wherein the reference input or command is constant for long periods of time, generally for the entire time interval for which the system is operational.

Stability

A system is stable when its impulse response approaches

zero as time approches infinity, or a system is stable if every bounded input produces a bounded output.

Routh Stability Criterion

The Routh stability criterion is a method for determining system stability that can be applied to an nth order characteristic equation of the form

$$a_n s^n + a_{n-1} s^{n-1} + \dots\dots + a_1 s + a_0 = 0$$

The criterion is applied through the use of a Routh table defined as under :

s^n	a_n	a_{n-2}	a_{n-4}	...
s^{n-1}	a_{n-1}	a_{n-2}	a_{n-5}	...
.	b_1	b_2	b_3	...
.	c_1	c_2	c_3	...
	......	....	....	

where a_n, a_{n-1}, ... a_0 are the coefficients of the characteristics equation and

$$b_1 = \frac{a_{n-1}\, b_{n-2} - a_n\, a_{n-3}}{a_{n-1}}$$

$$b_2 = \frac{a_{n-1}\, a_{n-4} - a_n\, a_{n-5}}{a_{n-1}} \quad \text{etc.}$$

$$c_1 = \frac{b_1 a_{n-3} - a_{n-1} b_2}{b_1}$$

$$c_2 = \frac{b_1 a_{n-5} - a_{n-1} b_3}{b_1} \quad \text{etc.}$$

The table is continued horizontally and vertically until only zeros are obtained. Any row can be multiplied by a constant before the next row is computed without disturbing the properties of the table. All the roots of this characteristic equation have negative real parts if and only if the elements of the first column of the Routh table have the same sign. Otherwise, the number of roots with positive real parts is equal to the number of changes.

Hurwitz Stability Criterion

The Hurwitz stability criterion is another method for determining whether or not all the roots of a characteristic equation have negative real parts. This criterion is applied through the use of determinants formed from the coefficients of the characteristic equation. It is assumed that the first coefficient, a_n, is positive. The determinants Δ_i for i=1, 2, ..., n—1 are formed as the principle minor determinants of the derterminant

$$\Delta_n = \begin{vmatrix} a_{n-1} & a_{n-3} & \cdots & \begin{bmatrix} a_0 \text{ if n odd} \\ a_1 \text{ if n even} \end{bmatrix} & 0 \ldots\ldots 0 \\ a_n & a_{n-2} & \cdots & \begin{bmatrix} a_0 \text{ if n odd} \\ a_1 \text{ if n even} \end{bmatrix} & 0 \ldots\ldots 0 \\ 0 & a_{n-1} & a_{n-3} & \ldots\ldots\ldots\ldots\ldots\ldots & 0 \\ 0 & a_n & a_{n-2} & \ldots\ldots\ldots\ldots\ldots\ldots & 0 \\ \cdots & \cdots & \cdots & \cdots \;\; \cdots \;\; \cdots & \cdots \\ 0 & \cdots & \cdots & \cdots \;\; \cdots \;\; \cdots & a_0 \end{vmatrix}$$

The determinates are thus formed as follows:

$$\Delta_1 = a_{n-1}$$

$$\Delta_2 = \begin{vmatrix} a_{n-1} & a_{n-3} \\ a_n & a_{n-2} \end{vmatrix} = a_{n-1}\, a_{n-2} - an\, a_{n-3}$$

$$\Delta_3 = \begin{vmatrix} a_{n-1} & a_{n-3} & a_{n-5} \\ a_n & a_{n-2} & a_{n-4} \\ 0 & a_{n-1} & a_{n-3} \end{vmatrix} = a_{n-1} a_{n-2} a_{n-3} + a_n a_{n-1} a_{n-5} - - a_n a_{n-3}^2 - a_{n-4} a_{n-5}$$

and so on upto Δ_{n-1}

All the roots of the characteristic equation have negative real parts if and only if

$\Delta_i > 0$ for i = 1, 2, ..., n.

Transfer Function

P(s) of a system is defined as that factor in the equation Y(s) multiplying that trans form of the input X (s).

Signal Flow Graph

A signal flow graph is a pictorial representation of the simultaneous equations describing a system. It graphically displays the transmission of signals through the system, as does the block diagram.

Methods of Analysis

(i) Nyquist

(ii) Root Locus analysis

(iii) Bode analysis

(iv) Nichols' chart analysis.

A Servo-mechanism

It is such a feedback control system wherein the controlled output is mechanical position on time derivation of mechanical position such as velocity or acceleration.

Linear Feed-back Control System

Any system whose input-output relationship can be described by a linear differential equation of the form

$$\sum_{i=0}^{M} a_i(t) \frac{d^i y}{dt^i} = \sum_{i=0}^{N} b_i(t) \frac{d^i x}{dt^i}$$

is linear. The principle of superposition is also valid in linear feedback control systems.

A Casual System or a Physically Realisable System

It is the one in which the output at any time (t) depends only on the present and the past inputs and that future inputs cannot be anticipated.

Free Response of a Control System

It is the one where the output explicitly depends upon the initial conditions i.e., input to the system is zero.

Forced Response of a Control System

It is the one where the output explicitly depends upon the input i.e. all initial conditions are assumed to be zero.

Steady state response of a Control System

It is that part of the response which does not approach zero when the time approaches infinity. On the other hand, transient response does approach zero with time to infinity.

Steady state response + Transient response = Total response
Free response + Forced response = Total response.

Impulse-response of a Control System

It is the most important of all. One of the reasons for this is unit impulse-response of any control system is nothing but its transfer function.

General form of a control system represented by a second order differential equation is as under

$$\frac{d^2y}{dt^2} + 2\tau w_n \left(\frac{dy}{dt}\right) + w_n^2 y = w_n^2 x$$

where τ = Damping ratio

w_n = Undamped natural frequency

Damping co-efficient, $\alpha = \zeta w_n$

Damped natural frequency $w_d = w_n \sqrt{1-\zeta^2}$

$$\text{Time constant } T = \frac{1}{\zeta w_n}$$

Laplace transform F (s) of any time domain function f(t):
It is given by

$$F(s) = \int_{0+}^{\infty} f(t)\, e^{-st}\, dt$$

(where $s = \sigma + jw$)

Inverse Laplace Transform : It is given by

$$\mathcal{L}^{-1} F(s) = \frac{1}{2\pi j} \int_{c-j\infty}^{c+j\infty} F(s)e^{st}\, ds$$

Properties of Laplace transform : These are as follows :

(a) $\mathcal{L}[Af_1(t) + Bf_2(t)] = AF_1(s) + BF_2(s)$

Conversely, $\mathcal{L}^{-1}[AF_1(s) + BF_2(s)] = Af_1(t) + Bf_2(t)$

A and B here are arbitrary constants.

(b) $\mathcal{L}\left[\int_0^t f(\tau)\, d\tau\right] = \frac{F(s)}{S}$

(c) $\mathcal{L}\left[\frac{d}{dt} f(t)\right] SF(s) - f(O^+)$

where $f(0^+)$ is the value of F(s) at $S = 0^+$

(d) According to Initial value theorem :

$$f(0^+) = \lim_{t \to 0} f(t) = \lim_{s \to \infty} SF(s)$$

According to Final value theorem :

$$f(\propto) = \lim_{t \to \infty} f(t) \lim_{s \to 0} SF(s)$$

(e) $\mathcal{L}\left[f\left(\frac{t}{a}\right)\right] = aF(as)$

$$\mathcal{L}^{1}\ F\left(\frac{S}{t}\right) = af(at)$$

(f) $\mathcal{L}[(t - T)] = e^{-sT} F(S)$

where $f(t - T) = 0$ for $t \leq T$

(g) $\mathcal{L}[e^{-at} f(t)] = F(S + a)$

(h) $\mathcal{L}[f_1(t) f_2(t)] = \frac{1}{2\pi j} \int_{c = j\infty}^{c + j\infty} F_1(w) F_2(s - w)\, dw$

$$\mathcal{L}^{-1}[F_1(s) F_2(s) = \int_{0+}^{t} f_1(\tau) f_2(t - \tau)\, d\tau$$

$$= \int_{0+}^{t} f_2(\tau) f_1(t - \tau)\, d\tau$$

Transfer Function of a Control System

It is defined as

$$P(S) = \frac{Y(S)}{X(S)}$$

where $Y(S) = \mathcal{L}y(t)$ and $X(S) = \mathcal{L}x(t)$.

Error Constants

Mathematically,

Position error constant, $k_p = \lim_{S \to 0} G(S)$

Velocity error constant, $k_s = \lim_{S \to 0} SG(S)$

Acceleration error constant, $k_a = \lim_{S \to 0} S^2G(S)$

Sensitivity of a Control System

It is the variation in system's transfer function with variation in one of its parameters.

Mathematically, $S^{T(k)}_k = \frac{dT(k)}{dk} \left[\frac{k}{T(k)} \right]$

where $T(k)$ = transfer function

k = any parameter

$S_k^{T(k)}$ = Sensitivity of $T(k)$ with respect to k.

Also $S^{T(k)}_k = \frac{k}{|T(k)|} \frac{d|T(k)|}{dk}$

and $S_k^{\phi T} = \frac{k}{\phi T} \left(\frac{d\phi S}{dk} \right)$

where $T(k) = |T(k)| e^{i\phi T}$

$|T(k)|$ = Magnitude of $T(k)$

ϕT = Phase of $T(k)$

$S_k^{|T(k)|}$ = Sensitivity of $|T(k)|$ w.r.t.(k)

$S_k^{\phi T}$ = Sensitivity of ϕT w.r.t.(k).

OBJECTIVE TYPE QUESTIONS

Q. 1.Under force-voltage analogy, velocity is analogous to which of the following ?

(A) Charge (B) Capacitance

(c) Inductance (D) Current.

Q. 2. In force-current analogy, capacitance is analogous to which of the following ?

(A) Mass (B) Velocity

(C) Displacement (D) Momentum.

Q. 3. The Laplace transform of e^{-2t} is given by

(A) $\frac{1}{2s}$ (B) $\frac{2}{s}$

(C) $\frac{2}{s+1}$ (D) $\frac{1}{s+2}$.

Q. 4. The increase Laplace transform of

$\frac{1}{s(s+2)}$ is given by

(A) $\frac{1}{2}[1 - 4^{2t}]$ (B) $\frac{1}{2}[1 + e^{-2t}]$

(C) $\frac{1}{2}[1 - e^{-2t}]$ (D) $\frac{1}{2}[1 + e^{2t}]$.

Q. 5. Laplace transform method of solution is applicable to equations containing

(A) Differential terms only

(B) Integral terms only

(C) Scalar terms only

(D) None of the above.

Q. 6. If a zero appears on the first column of the Routh table, the system is

(A) Necessarily stable

(B) Necessarily unstable

(C) Marginally stable

(D) None of these.

Q. 7. Which of the following informations can be concluded about the transient behaviour of the system?

(A) 25% maximum overshoot to step command

(B) 50% maximum overshoot to step command

(C) 75% maximum overshoot to step command

(D) None of the above.

Q. 8. The d.c. gain of the system represented by the transfer function

$$P(s) = \frac{1}{s+1} \text{ is equal to}$$

(A) 1 (B) 2

(C) 5 (D) 10.

Q. 9. The transfer function of a system is given by

$$P(s) = \frac{2}{s+2}$$

for ω=10, the gain will be equal to

(A) 0.444 (B) 0.333

(C) 0.25 (D) 0.196.

Q. 10. In Fig. 11.1 the ramp test signal is represented by

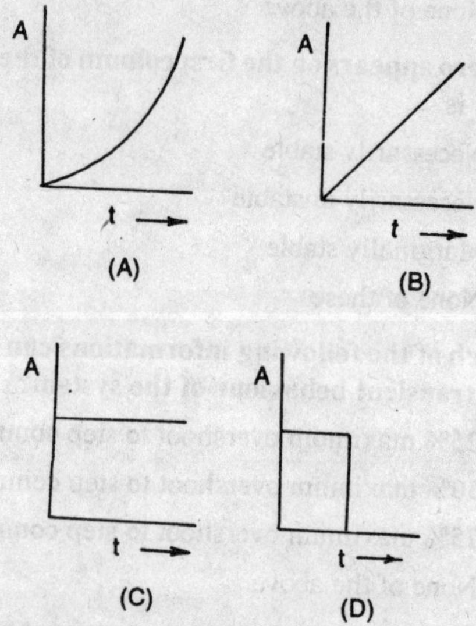

Fig. 11.1.

(A) Figure A (B) Figure B
(C) Figure C (D) None of the figures.

Q. 11. Damping is proportional to

(A) gain (B) $\frac{1}{\text{gain}}$
(C) $\sqrt{\text{gain}}$ (D) $\frac{1}{\sqrt{\text{gain}}}$.

Q. 12. A type 1 system under parabolic input will have

(A) Actuating signal which will increase with time
(B) Actuating signal which will decrease with time
(C) Parabolic output
(D) Any of the above.

Q. 13. Which of the following best defines a transfer function?

(A) Ration of system response to system input function

(B) Ratio of system input function to system response.

(C) Laplace transform of system response minus the Laplace transform of the system input function

(D) Ratio of the Laplace transform of the system response to the Laplace transform of the system input function.

Q. 14. Analysis of control systems by Laplace transform technique is NOT possible for which one of the following ?

(A) Linear systems

(B) Time-invariant systems

(C) Discrete-time systems

(D) Unstable continuous-time systems.

Q. 15. Suppose an R-L-C electrical network with supply voltage u(t) and output voltage y(t) is described by

$$\frac{d^2y}{dt^2} + 6\frac{dy}{dt} + 25\, y(t) = 50\, u(t)$$

If sinusodial input voltage with peak amplitude 150 volts and frequency 60 Hz is applied to this network,what is the frequency of the sinusodial steady-state output ?

(A) The system's undamped natural frequency w_n

(B) The system's damped natural frequency $w_n \sqrt{1 - \epsilon^2}$

(C) The input frequency 60 Hz

(D) Somewhat less than 60 Hz.

Q. 16. The unit step response of a linear time-invariant system can be obtained by

(A) Inverse Laplace transform of the transfer function

(B) Integration of the impulse response

(C) Using the Routh-Hurwitz criterion

(D) Applying the Laplace final value theorem.

Q. 17. Inverse Laplace transform of F (s) is given by which of the following ?

(A) $\mathcal{L}^{-1}F(s) = \frac{1}{2\pi i} \int_{c-j\infty}^{c+j\infty} F(s)e^{st}\, ds$

(B) $\mathcal{L}^{-1} F(s) = \int_{-\infty}^{\infty} F(s)\, e^{st}\, ds$

(C) $\mathcal{L}^{-1} F(s) = \frac{1}{2\pi j} \int_{-\infty}^{\infty} F(s)e^{st}\, ds$

(D) None of these.

Q. 18. If $F_1(s)$ and $F_2(s)$ are the laplace transforms of $f_1(t)$ and $f_2(t)$ respectively, then laplace transform of a function {A f_1 (t) + Bf_2 (t) } will be equal to

(A) $AF_1(s) + BF_2(s)$

(B) $\frac{F_1(s)}{A} + \frac{F_2(s)}{B}$

(C) $BF_1(s) + AF_2(s)$

(D) None of these.

Q. 19. What is the transfer function of a system having a gain factor of 6 and pole-zero map as shown in Fig. 11.2.

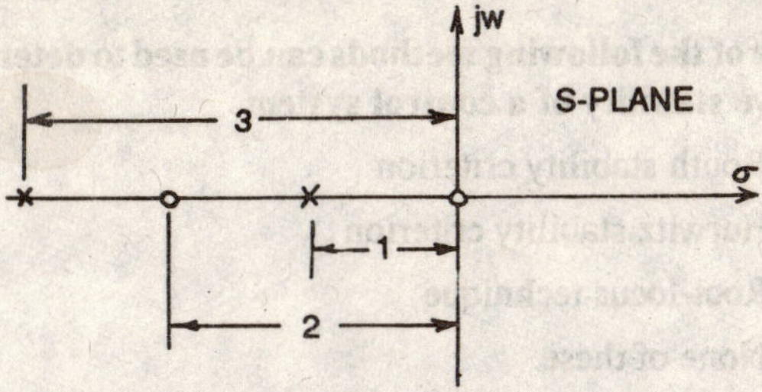

Fig. 11.2.

(A) $\dfrac{6(S+1)(S+3)}{(S+2)S}$ (B) $\dfrac{6S(S-2)}{(S+1)(S+3)}$

(C) $\dfrac{6S(S+2)}{(S+1)(S+3)}$ (D) None of these.

Q. 20. An impulse when applied to a certain system resulted in an exponential output of e^{-3t}. The transfer function of this system will be

(A) Cannot be determined from given data

(B) is $\dfrac{1}{(S+3)}$ (C) is $\dfrac{1}{(S-3)}$

(D) is $\dfrac{3}{(S+3)}$.

Q. 21. In order to have the impulse response of a control system approaching zero with time tending to infinity

(A) The poles if the system must have position real parts,

(B) The poles of the system must lie on the L.H.S. of the S-plane,

(C) The zero of the system must be on the L.H.S. of the S-plane,

(D) None of these.

Q. 22. One of the following methods can be used to determine the relative stability of a control system

(A) Routh stability criterion

(B) Hurwitz stability criterion

(C) Root-locus technique

(D) None of these.

Q. 23. Routh table was constructed from the characteristic equation of a control system. The first column of the table contained the following integers, 1, 4, 5, 6, 13. The system is

(A) Stable (B) Unstable

(C) Marginally stable (D) None of these.

Q. 24. Figure 11.3 shows a Niquist plot. The Gain Margin is

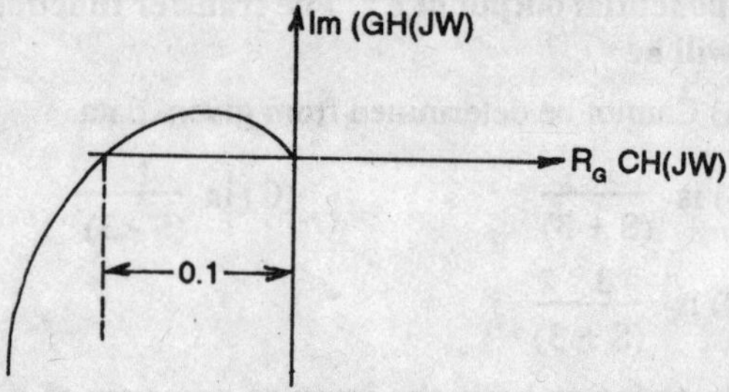

Fig. 11.3.

(A) 0.1 (B) 10

(C) 100 (D) None of these.

Q. 25. Gain Margin can be determined from the root-lows plot by

(A) $\dfrac{\text{Value of K at imaginary crossover}}{\text{Design value of K}}$

(B) $\dfrac{\text{Design value of K}}{\text{Value of K at imaginary crossover}}$

(C) $\dfrac{\text{Design value of K}}{\text{Value of K at real axis crossover}}$

(D) $\dfrac{\text{Value of K at real axis crossover}}{\text{Design value of K}}$

Q. 26. Figure 11.4 shows the real axis portion of the root-locus of a certain system. This portion shown-tells us something about the gain factor 'k' of the system will be given by

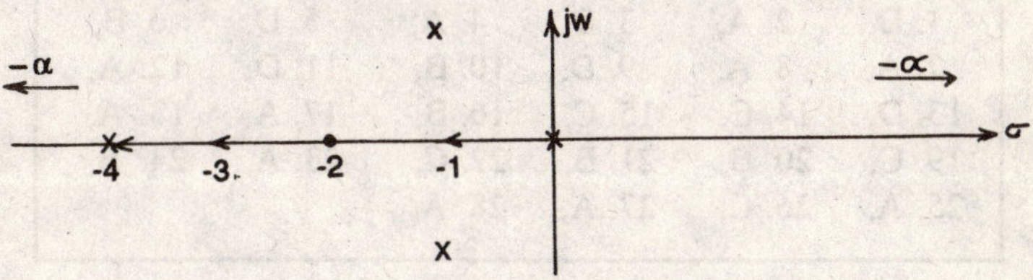

Fig. 11.4.

(A) K = 0 (B) K = 1

(C) K > 0 (D) K < 0.

Q. 27. 'Centre of asymptotes' is a very important factor in drawing the root-locus for a control system whose open loop transfer function is known. What is it for a system with open loop transfer function as

$$\frac{3(s+2)}{s^2(s+4)}$$

(A) -1 (B) - 2

(C) - 3 (D) 0.

Q. 28. The angles made by asymptotes in a root-locus plot are given by (assume gain factor to be greater than zero).

(A) $\dfrac{(2l+1)\ 180}{n - m}$ degrees

(B) $\dfrac{(2l)\ 180}{n - m}$ degrees

(C) (2l+1) 180 degrees

(D) (2l) 180 degrees.

ANSWERS

1. D,	2. A,	3. D,	4. A,	5. D,	6. B,
7. A,	8. A,	9. D,	10. B,	11. D,	12. A,
13. D,	14. C,	15. C,	16. B,	17. A,	18. A,
19. C,	20. B,	21. B,	22. C,	23. A,	24. B,
25. A,	26. C,	27. A,	28. A.		

Chapter 12

Power Electronics

STUDY MATERIAL

dc-dc-convertor

If the output of an invertor circuit is rectified and filtered, it becomes a dc-dc-convertor

Invertor

It is a power conversion device that converts applied d.c. input into a.c. output.

Classes of Invertors and Convertors

They are (i) Ringing choke or Flyback invertors/convertors (ii) Forward invertors/convertors.

Flyback Invertors/Convertors

In them the power is transferred to the load only when the switch is in OFF state (If a switch is a transistor, the power will be transferred only when the transistor is in the cut off region).

Forward Convertors

In them the power is continuously transferred irrespective of the state of the switching device.

The design equations of a flyback convertor

(i) The average value of the current in the primary winding

of the transformer is given by

$$I_P(av) = \frac{P_{out}}{\eta V_{IN}}$$

where P_{out} = output power

η = convertor efficiency

V_{IN} = Input voltage

The peak primary current ($\hat{I}_P$)

It can be expressed in terms of the average current and subsequently other parameters as

$$\hat{I}_P = \left(\frac{P_{out}}{\eta V_{in}}\right)\left(\frac{2T}{t_{on}}\right)$$

Where T = 1/f : where f = frequency of operation

t_{on} = ON-time

Also $\frac{t_{on}}{T}$ = Duty cycle.

For a 50% duty cycle

$$\hat{I}_P = \left(\frac{4P_{out}}{\eta V_{in}}\right)$$

(ii) The primary inductance (L_P)

It is given by

$$L_P = \left(\frac{\eta V_{in}^2}{2P_{out}\, f}\right) \times \left[\frac{\left(\frac{N_P}{N_S}\right)^2 V_{out}^2}{V_{IN}^2 \left(\frac{N_P}{N_S}\right)^2 + V_{out}^2 + 2V_{IN}\, V_{out}\, \frac{N_P}{N_S}}\right]$$

Where N_P = No. of primary turns

N_S = No. of secondary turn

(L_P) is also given by

$$L_P = \left(\frac{\eta Vin^2}{2Pout}\right)\left(\frac{t_{on}}{T}\right)^2\left(\frac{1}{f}\right)$$

(iii) The selected transistor should have V_{CB} (max) which is atleast equal to

$$\left[\ V_{in}^{\ 2} + \left(\frac{N_P}{N_S}\right)\ Vout\ \right]$$

It should have a value which is slightly greater than the one given by the above expression.

(iv) The value of output capacitor is decided by the criterion that the time constant $R_L C_{out}$ (where R_L is the load resistance) should atleast be ten times the on-time (t_{on}).

Mathematically

$$Cout = \frac{(10\ ton\ Pout)}{Vout^2}$$

The design equations of a push-pull type forward convertor

The convertor employs two transistors and one saturable type transformer. It is self-oscillation.

(i) The average value of the collector current in the conducting transistor is given by

$$I_C\ (av) = \frac{Pout}{\eta V_{in}}$$

The peak value of the collector current in the conducing transistor

It is given by

$$\hat{I}_P = \frac{V_{in}}{4f L_P}$$

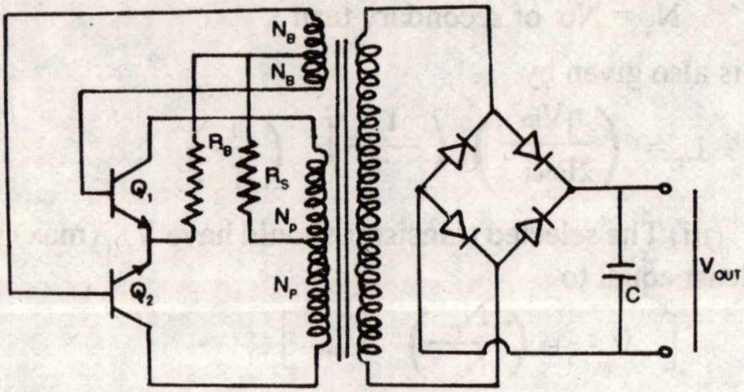

Fig. 12.1.

where f = frequency of operation

L_P = Primary inductance.

V_{IN} = Input voltage

η = Efficiency.

The primary current can then be expressed by

$$I_{P(max)} = \left(\frac{V_{in}}{4f L_P}\right) + \left(\frac{P_{out}}{\eta V_{in}}\right)$$

(ii) Primary inductance (L_p):

It is expressed by

$$L_P = (N_P \, B_{max} \, A \times 10^{-8}) / \hat{I}_P$$

where B_{max} = Maximum saturation flux denisity

A = Area of cross-section.

(iii) No. of primary turns can be calculated from

$$V_{IN} = (4 \, N_P \, B_{max} \, A \times 10^{-8})$$

where B_{max} is in Gauss

A is in CM^2

(iv) The No. of turns in the feedback winding

It is given by

$$N_B = N_P\left(\frac{2V_{BE(max)}}{V_{in}}\right)$$

$V_{BE(max)}$ = Maximum base to emitter voltage of the transistor

(v) (R_S) and (R_B) can be calculated from

$$R_S = \left(\frac{V_{in} - 0.6}{0.6}\right) R_S$$

This relationship ensures a voltage of 0.6 V at the centre tap of the feadback winding when the supply voltage is applied.

(vi) The value of output capacitor

It is given by

$$Cout = \left(\frac{100}{R_L f}\right)$$

(vii) The selected transistor should have a $V_{CB(max)}$ rating which is greater than twice the applied input voltage. Precisely,

$$V_{in(max)} < \left(\frac{V_{CB(max)} - V_{BE(wind)}}{2}\right)$$

where $V_{BE(wind)}$ is the induced voltage in one half of the base winding.

Choppers can be divided into three categories depending upon the possibility of reversing the output voltage or output current or both.

(i) A type choppers are not capable of changing the direction of either the output voltage or output current.

(ii) B-type choppers are capable of changing the direction of output current. However, the direction of output voltage cannot be change.

(iii) In C-type choppers, both the alterations are possible.

(iv) A-type chopper is a I-Quadrant chopper.

B-type chopper is a II-Quadrant chopper.

C-type chopper is a IV-Quadrant chopper.

(i) SCRS and triacs can be used to control the amount of power delivered to the load. By controlling the amount of time for which an SCR or a triac is kept 'ON' the amount of current delivered to the load can be varied.

(ii) A single SCR can be used to have a half wave power control whereas triac can achieve power control in both half cycles. Figure 12.2 shows the basic circuit arrangement of a half wave power control with one SCR Fig. 12.2 (b) shows the input voltage and load voltage waveforms. In these waverforms, the

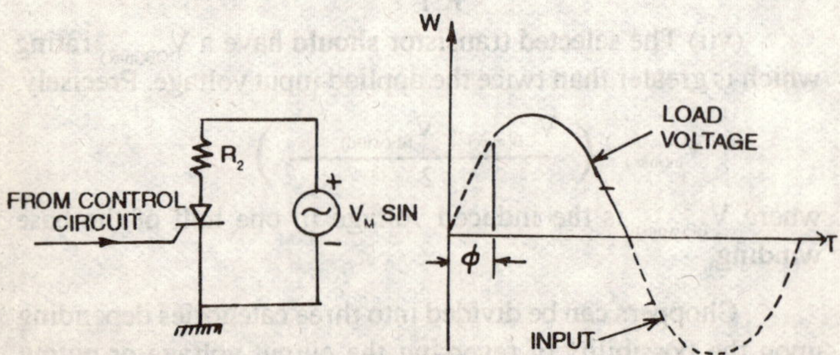

Fig. 12.2.

holding voltage (V_H) of SCR has been neglected. The load current in such a case is given by

(When SCR is ON)

$$I_L = \left(\frac{V_M \sin wt - V_H}{R_L}\right)$$

But the SCR is 'ON' only for a time equivalent to $(\pi - \phi)$radians, where ϕ is the conduction angle i.e., the angle where the SCR turns ON. The average current is thus given by

$$I_{AV} = \frac{V_M}{2\pi R_L}\left(1 + \cos\phi\right)$$

The above expression clearly shows that the higher value of results in reduced I_{AV} whereas a lower value of ϕ gives increased I_{AV}. Thus (I_{AV}) can be controlled by controlling the conduction angle (ϕ).

(iii) Figs. 12.3 (a) and (b) show the use of a triac to control a.c. power in both the half cycles and the waveform of applied voltage and load voltage. Figure shows two SCRS in inverse parallel instead of triac.

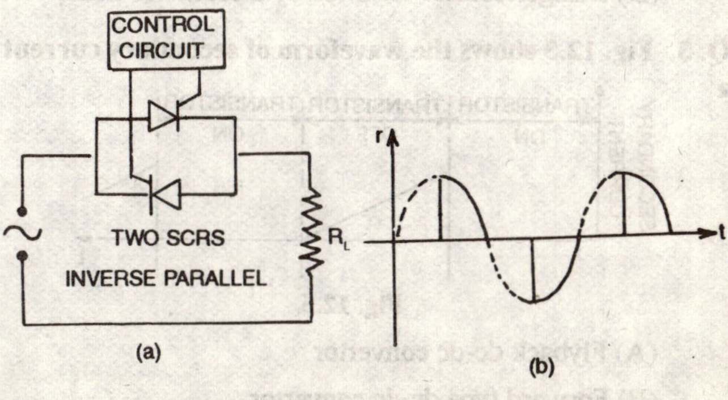

Fig. 12.3.

OBJECTIVE TYPE QUESTIONS

Q. 1. The output of an invertor circuit is

(A) Always a d.c. signal

(B) Always sinusoidal in shape

(C) Always a square waveform

(D) Always a.c. in nature.

Q. 2. Fig. 12.4 shows a dc-dc convertor in the block schematic form. Tell, what unnamed block is?

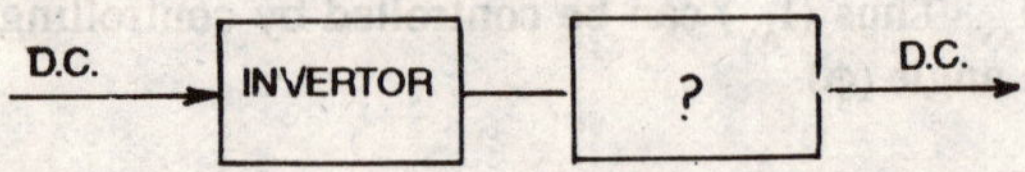

Fig. 12.4.

(A) Rectifier and filter (B) Regulator

(C) Filter

(D) Bridge rectifier followed by a capacitor filter.

Q. 3. Fig. 12.5 shows the waveform of secondary current of a

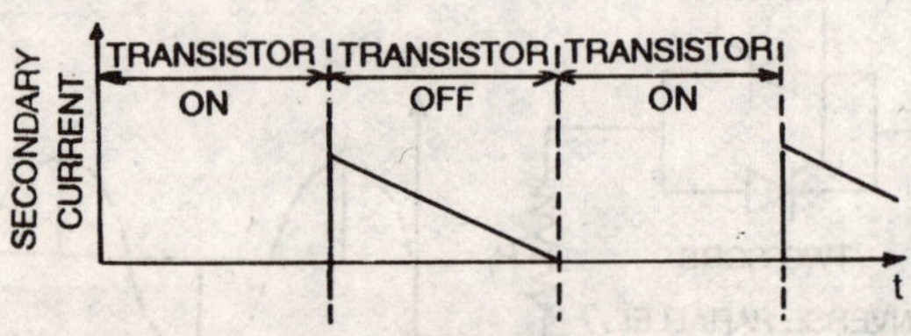

Fig. 12.5.

(A) Flyback dc-dc convertor

(B) Forward type dc-dc convertor

(C) Push-pull type dc-dc convertor

(D) None of these.

Q. 4. Recognize the circuit shown in Fig. 12.6 and close the correct answer:

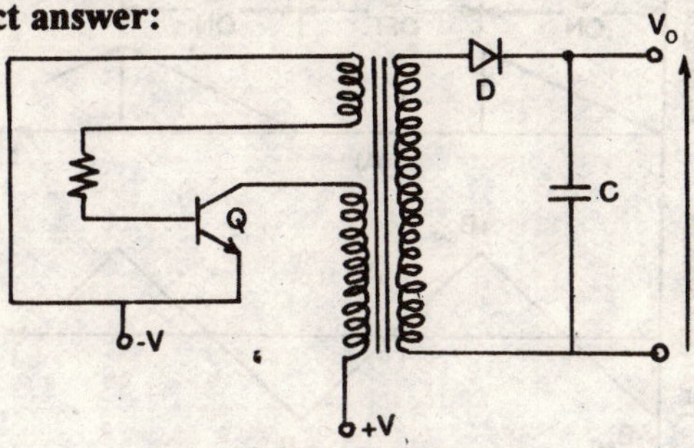

Fig. 12.6.

(A) Ringing choke convertor

(B) Forward type convertor.

(C) Tuned amplifier.

(D) Flyback invertor.

Q. 5. The number of primary turns needed to get the desired primary inductance on a core having an air gap of (l_a), permeability of μe magnetic path length, (l_e) and area of cross-section (A_e) can be calculated by using

(A) $L = \dfrac{N_P^2 \, Ae}{\dfrac{le}{\mu e} + la}$ (B) $L = \dfrac{\mu e \, N_P^2 \, Ae}{le + la}$

(C) $L = \dfrac{\mu e \, N_P^2 \, Ae}{le}$ (D) $L = \dfrac{\mu e \, N_P^2 \, Ae}{la}$

Q. 6. Figure 4 (A) shows the primary current waveform of which of the following?

(A) Flyback convertor.

(B) Forward convertor.

(C) Ringing choke invertor.

(D) none of these.

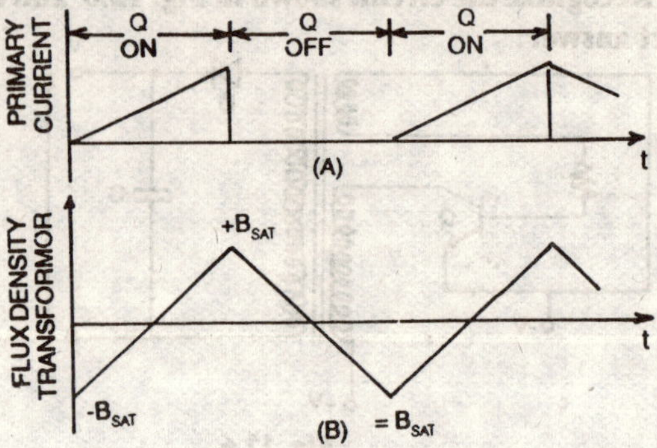

Fig. 12.7.

Q. 7. Figure 12.7 shows the flux density waveform of which of the following?

(A) A flyback dc-dc convertor.

(B) A push pull convertor.

(C) A forward convertor.

(D) None of these.

Q. 8. See the circuit shown in Fig. 12.8 and select the correct answer:

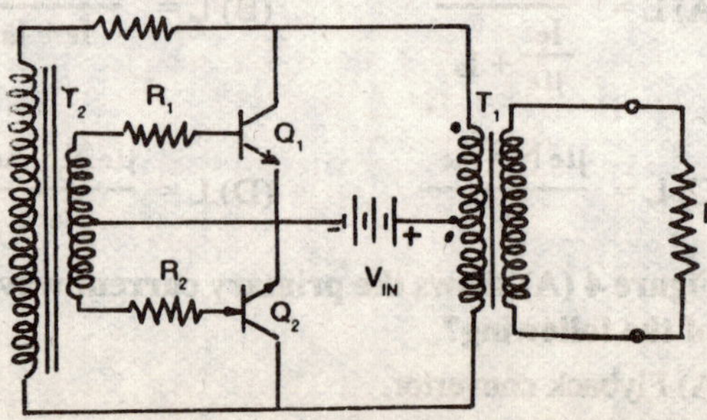

Fig. 12.8.

(C) Push-pull invertor.

(B) Forward convertor.

(C) Ringing choke invertor.

(D) Ringing choke convertor.

Q. 9. Transformer (T1) shown in Fig. 12.8 is of which type?

(A) Non-saturating type (B) Saturating type

(C) Neither of the two.

Q. 10. See the circuit shown in Fig. 12.9. It is an a.c. power control circuit. A sinusoidal waveform when applied at the input of the circuit will result in an output waveform given by which of the following?

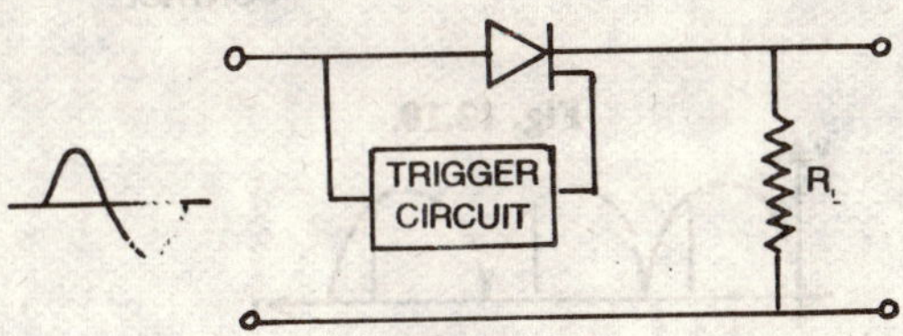

Fig. 12.9.

(A) 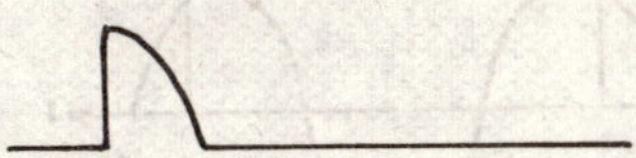

(B) 

(C) 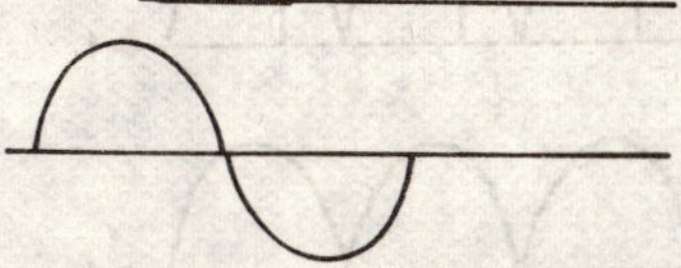

(D) 

Q. 11. See the circuit shown in Fig. 12.10. If a sinusoidal voltage is applied at the input. The load voltage will be given by which of the following?

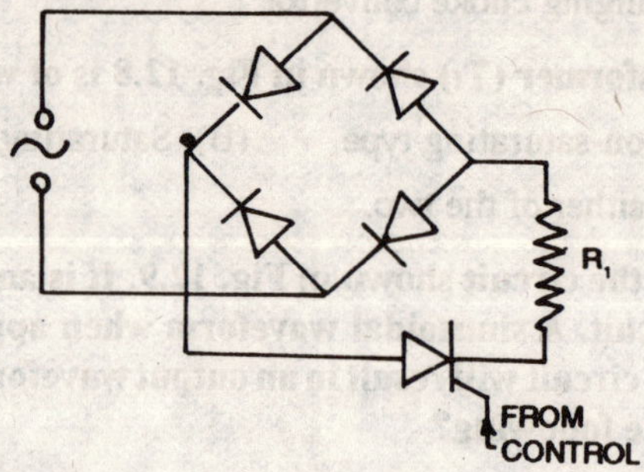

Fig. 12.10.

(A) 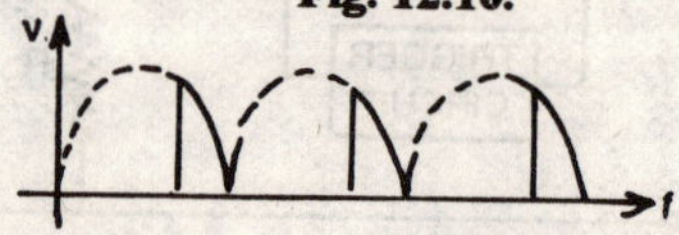

(B) 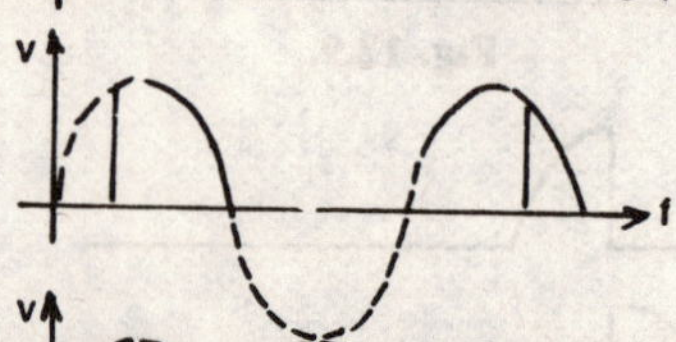

(C) 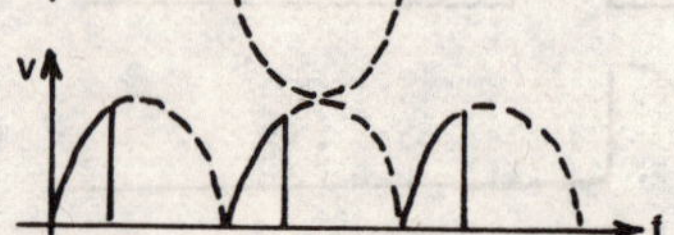

(D) 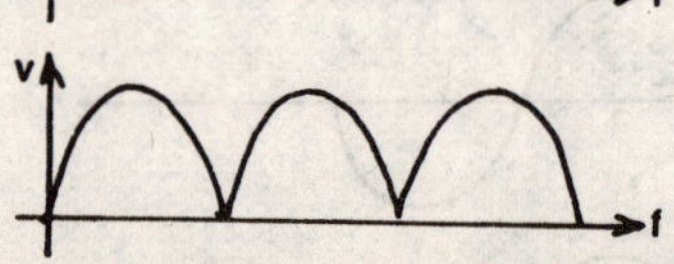

Q. 12. The maximum value of primary current that can flow in a push-pull convertor when the transistor is conducting can be expressed by which of the following?

(A) $\left\{ \frac{V_{out}}{\eta V_{in}} + \frac{V_{in}}{4f L_P} \right\}$ (B) $\frac{V_{out}}{\eta V_{in}}$

(C) $\frac{V_{in}}{4f L_P}$ (D) None of these.

Q. 13. Which of the following convertor configuration has a higher efficiency?

(A) Push-pull convertor

(B) Flyback convertor

(C) Both have almost identical efficiencies.

Q. 14. Higher the frequency of operation of a convertor,

(A) Larger is the size and weight.

(B) Smaller is the size and weight.

(C) Higher is the efficiency.

(D) Lesser is the efficiency.

Q. 15. The no. of primary turns needed to get an inductance of (L) can be calculated from which of the following?

(A) $L_P = \left\{ \frac{N_P \, B_{max} \, A \times 10^{-8}}{\hat{I}_P} \right\}$ where $\hat{I}_P$ is the peak primary current

(B) $\left\{ \frac{N_P \, B_{max} \, A}{\hat{I}_P} = L_P \right\}$

(C) $\left\{ L_P = \frac{N^2_P \, B_{max} \, A}{I_P} \right\}$

(D) None of these.

ANSWERS

1. D,	2. A,	3. A,	4. A,	5. A,	6. C,
7. B,	8. A,	9. A,	10. A,	11. A,	12. A,
13. A,	14. B,	15. A.			

Chapter 13

Networks

STUDY MATERIAL

Relation between Instantaneous Voltage, Current, Power and Energy in case of Resistances:

$$v_R = R_{iR}$$

$$p = R_{iR}^{\ 2}$$

$$w = \int_0^t pdi$$

Expression for Capacitors and Inductors

$$q_C = Cv_c$$

$$i_c = \frac{Vdv_c}{dt}$$

$$p = v_c i_c \qquad \text{In case of capacitors}$$

$$w = \int_0^t pdt$$

$$v_L = L\frac{di_L}{dt}$$

$$i_L = \frac{I}{L}\int v_L dt$$

$$p = v_L i_L \qquad \text{In case of inductors}$$

$$w = \int_0^t pdt$$

The overall resistance in case of series and parallel connections of resistors is respectively given by

$$R_{tot} = R_1 + R_2 + R_3 + \ldots\ldots \text{ for series connection}$$

$$\frac{1}{R_{tot}} = \frac{1}{R_1} + \frac{1}{R_2} + \frac{1}{R_3} + \ldots\ldots \text{for parallel connection}$$

Similar expressions in case of capacitors and inductors are as under :

$$C_{tot} = C_1 + C_2 + C_3 + \ldots\ldots \text{for parallel connection}$$

$$\frac{1}{C_{tot}} = \frac{1}{C_1} + \frac{1}{C_2} + \frac{1}{C_3} + \ldots\ldots \text{ for series connection}$$

and $L_{tot} = L_1 + L_2 + L_3 + \ldots\ldots$ for series connection

$$\frac{1}{L_{tot}} = \frac{1}{L_1} + \frac{1}{L_2} + \frac{1}{L_3} + \ldots\ldots \text{for parallel connection.}$$

Kirchoff's Voltage Law (KVL)

The algebraic sum of voltage around any closed path of a network when the path is traversed in a single direction is zero.

Kirchoff's Current Law (KCL)

The algebraic sum of currents entering a node or junction always equals the sum of currents leaving the same junction.(i.e. the algebraic sum is zero).

These laws are valid for steady state a.c. analysis, transient analysis, steady state d.c. analysis.

For steady state a.c. analysis do the phasor addition.

For instantaneous analysis instantaneous voltage and currents are to be considered.

Some Important Tools for Network Analysis

(i) Star-Delta Conversion and vice-versa

(ii) Power transform theorem

(iii) Millman's theorem

(iv) Superposition theorem

(v) Reciprocity theorém

(vi) Compensation theorem

(vii) Thevenin's theorem

(viii) Norton's theorem.

Maximum Power Transform Theorem

Power transferred from a source to a load is maximum when the load impedence is complex conjugate of the source impedance.

Superposition Theorem

In a linear bilateral active network containing two or more than two sources, a branch current or node voltage in any part of the network can be determined by considering only one source at a time with all other sources removed and then superposing the results.

Reciprocity Theorem

In a linear, bilateral and active network with a single source the ratio of excitation to response remains unchanged when the two interchange their position.

Compensation (or Substitution) Theorem

If in a linear, active network, a current I flows through an impedance Z, then the change in current (Δ I) produce due to change in impedance (Δ Z) is given by the current that would flow if a compensation source equal in magnitude and phase to (I Δ Z) is connected in series with (Z) with all other sources removed.

Mill Man's Theorem

A number of voltage sources V_1, V_2, V_3 .... Vn with respective internal impedances Z_1, Z_2 .... V_n when connected in parallel are electrically equipment to a single voltage source V with internal impedance Z

$$\text{where} \quad V = \frac{\frac{V_1}{Z_1} + \frac{V_2}{Z_2} + \frac{V_3}{Z_3} + \ldots\ldots \frac{V_n}{Z_n}}{\left(\frac{1}{Z_1} + \frac{1}{Z_2} + \ldots\ldots \frac{1}{Z_n}\right)}$$

$$Z = \frac{1}{\left(\frac{1}{Z_1} + \frac{1}{Z_2} + \ldots\ldots \frac{1}{Z_n}\right)}$$

Similarly a number of current sources I_1, I_2, I_3 .... I_n with respective shunt impedances Z_1, Z_2, ... Z_n when connected in parallel are electrically equivalent to a single current source I with shunt impedance Z, where

$$I = I_1 + I_2 + I_3 + \ldots\ldots\ldots I_n$$

$$Z = \frac{1}{\left(\frac{1}{Z_1} + \frac{1}{Z_2} + \frac{1}{Z_3} + \ldots\ldots \frac{1}{Z_n}\right)}$$

Star-connected impedances Z_A, Z_B and Z_C respectively connected between points A, B and C and a common point can be converted into equivalent Delta connected impedances Z_{AB} (between point A and B), Z_{BC} (between points B and C) Z_{CA} (between points C and A).

$$Z_{AB} = \frac{Z_A Z_B + Z_B Z_C + Z_C Z_A}{Z_C}$$

$$Z_{BC} = \frac{Z_A Z_B + Z_B Z_C + Z_C Z_A}{Z_A}$$

$$Z_{CA} = \frac{Z_A Z_B + Z_B Z_C + Z_C Z_A}{Z_B}$$

Similarly, Delta connected impedances can be converted into equivalent star-connected impedances using following formula

$$Z_A = \frac{Z_{AB}\, Z_{CA}}{Z_{AB} + Z_{BC} + Z_{CA}}$$

$$Z_B = \frac{Z_{AB}\, Z_{BC}}{Z_{AB} + Z_{BC} + Z_{CA}}$$

$$Z_C = \frac{Z_{BC}\, Z_{CA}}{Z_{AB} + Z_{BC} + Z_{CA}}$$

Current through an inductance and voltage across a capacitance change instantaneously.

Growing current exponential in a series R-L network

(assuming zero initial conditions), It is given by

$$i = \frac{V}{R}\left(1 - e^{-\frac{R}{L}t}\right)$$

Decaying current exponential in a series R-L network

It is given by

$$i = \frac{V}{R}\, e^{-\frac{R}{L}t}$$

Growing voltage exponential in a series R-C network

(assuming zero initial conditions). It is given by

$$v = V\left(1 - e^{-t/RC}\right)$$

Decaying voltage exponential in series R-C network

It is given by

$$v = Ve^{-t/RC}$$

In a series R-L-C network, growing current transient has three possible solutions. The current transient is overdamped when $\frac{R}{2L} > \frac{1}{\sqrt{LC}}$. It is critically damped when $\frac{R}{2L} = \frac{1}{\sqrt{LC}}$ and finally it is underdamped when $\frac{R}{2L} < \frac{1}{\sqrt{LC}}$.

When two coils with self-inductances (L_1), and (L_2) are closely coupled with a coupling co-efficient of (k), *the magnitude of mutual inductance* (M) *between the two coils* is given by

$$M = k\sqrt{L_1 L_2}.$$

Quality facto (Q-factor)

It is defined as

$$Q = 2\pi\left\{\frac{\text{Maximum energy stored}}{\text{Energy dissipated per cycle}}\right\}$$

Q-factor for a practical inductor

It is given by

$$Q = \left(\frac{\omega L}{R}\right)$$

Q-factor for a practical capacitor

It is given by

$$Q = (\omega CR)$$

The Q-factor for a series R-L-C circuit

It is generally applied as resonance when

$$Q = \frac{\omega_0 L}{R} = \frac{1}{\omega_0 CR} = \frac{1}{R}\sqrt{\frac{L}{C}}$$

Q-factor of the parallel R-L-C network at resonance

It is given by

$$Q = \frac{R}{\omega_0 L} = \omega_0 RC = R\sqrt{\frac{C}{L}}$$

In Trignometric Fourien series expansion, a periodic function f(t) so that f (t) = f (t + T) is given by

$$f(t) = \frac{a_0}{2} + a_1 \cos \omega t + a_2 \cos 2\omega t + \ldots\ldots a_n \cos n\omega t$$

$$+ b_1 \sin \omega t + b_2 \sin 2\omega t + \ldots\ldots + b_n \sin n\omega t.$$

where $a_n = \frac{1}{\pi}\int_0^{2\pi} f(t) \cos n\omega t \, d(\omega t)$

$$b_n = \frac{1}{\pi}\int_0^{2\pi} f(t) \sin n\omega t \, d(\omega t)$$

$$\frac{a_0}{2} = \text{Average value of the waveform.}$$

Function f(t) can also be expressed in exponential series as

$$f(t) = \ldots + A_{-2} e^{-j2\omega t} + A_{-1} e^{-j\omega t} + A_0 + A_1 e^{j\omega t} + A_2 e^{-j2\omega t} + \ldots\ldots$$

where $A_n = \frac{1}{T}\int_0^T f(t) e^{-jn\omega t} dt = \frac{1}{2\pi}\int_0^{2\pi} f(t) e^{-jn\omega t} d(\omega t)$

For a function f(t) to be even,

$$f(t) = f(-t)$$

for a function f(t) to be odd,

$$f(t) = -f(-t)$$

For a function f(t) to be half-wave symmetric

$$f(t) = -f(t + T/2)$$

OBJECTIVE TYPE QUESTIONS

Q.1. Calculate the value of (Y) resistance in the Figure 13.1.

(A) 4Ω (B) 1Ω

(C) 10Ω

(D) $\frac{50}{(15 + X)}$ and its value cannot be determined as (x) is unknown

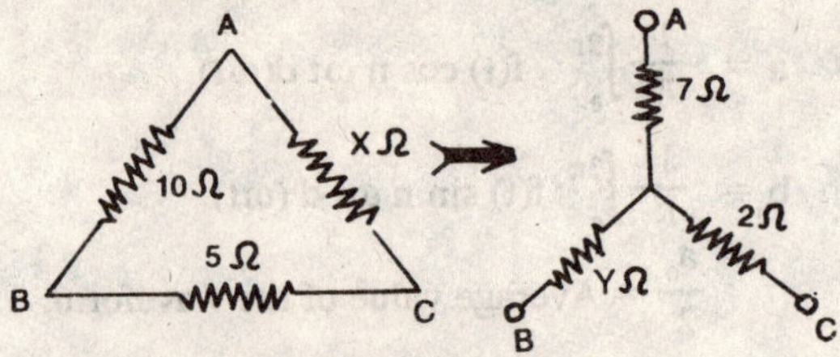

Fig. 13.1.

Q.2. Using the network given in Fig. 13.1 find the value of resistance (x) Ω

(A) Given by $\left\{ \frac{50}{15 + x} = Y \right\}$

(B) is 35Ω (C) is 4Ω

(D) Cannot be determine from available data.

Q.3. If all three elements of a given DELTA-connected network are pure inductive reactances, the three elements of the corresponding star connected network

(A) Will all be pure capacitive reactances

(B) Will all be pure inductive reactances

(C) Can be either inductive or capacitive reactances

(D) Will be inductive or capacitive reactances depending upon the Delta connected inductive reactances magnitudes.

Q.4. The resistances R_A, R_B and R_C have been connected in a STAR fashion as shown in Fig. 13.2. The equivalent DELTA resistance R_{AB} is given by

(A) $R_A + R_B + \frac{R_A R_B}{R_C}$ (B) $R_B + R_C + \frac{R_B R_C}{R_A}$

(C) $\frac{R_A R_B}{R_A + R_B + R_C}$ (D) $\frac{R_A R_B + R_B R_C + R_C R_A}{R_C}$

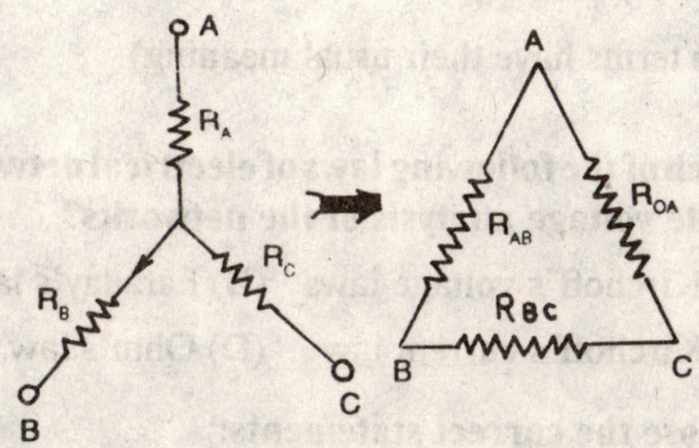

Fig. 13.2.

Q.5 For Fig. 13.2, calculate the value of resistance R_{CA}

(A) $R_A + R_C + \frac{R_A R_C}{R_B}$ (B) $R_A + R_B + \frac{R_A R_B}{R_C}$

(C) $\dfrac{R_A R_C}{R_A + R_B + R_C}$ (D) $\dfrac{(R_A + R_B + R_C)}{R_A R_C}$

Q.6. For the equivalent Star-Delta networks of Fig. 13.2, how can you determine the resistance R_{BC} from the knowledge of R_{AB}?

(A) $R_{BC} = \dfrac{R_{AB} \times R_A}{R_C}$ (B) $R_{BC} = \dfrac{R_{AB} R_C}{R_A}$

(C) $R_{BC} = \dfrac{R_{AB}(R_A R_B)}{(R_A + R_B)}$

(D) R_{BC} can be determined only if R_A, R_B and R_C are known.

Q.7. The Driving point impedance of a network can be calculated from

(A) $\dfrac{\Delta_Z}{\Delta_{11}}$ (B) $\dfrac{\Delta_Y}{\Delta_{11}}$

(C) $\dfrac{\Delta_{11}}{\Delta_Z}$ (D) $\dfrac{\Delta_{11}}{\Delta_Y}$

(The terms have their usual meaning).

Q.8. Which of the following laws of electrical network is used in the Node voltage analysis of the networks?

(A) Kirchoff's voltage laws (B) Faraday's laws

(C) Kirchoff's current law (D) Ohm's law.

Q.9. Choose the correct statements:

(A) In the Mesh current analysis, the directions of the loop currents should be so chosen that no branch should have more than two currents otherwise the result are likely to be incorrect

(B) In the Node voltage analysis, if outgoing currents are taken as positive, then incoming currents must be taken as negative and vice versa

(C) In the Node voltage analysis, all outgoing current must always be taken as negative and all incoming currents must always be taken as positive

(D) In Mesh current analysis, all loop currents should either be clockwise or anticlockwise.

Q.10. In the network of Fig. 13.3, in case of maximum power transfer, the power delivered to the load resistance (R_L) is

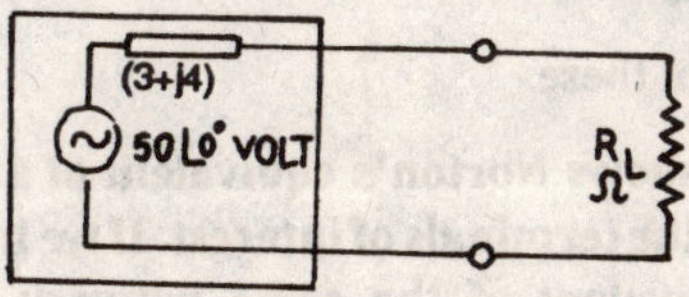

Fig. 13.3.

(A) 125 watts (B) 156 watts

(C) 110 watts

(D) Cannot be determined from given data.

Q.11. In the network of Fig. 13.3, for maximum power transfer what is magnitude of R_L ?

(A) 3Ω (B) 4Ω

(C) 7Ω (D) 5Ω.

Q.12. In the network of Fig. 13.3, if the active network's equivalent impedance was only 3Ω instead of (3+j4)Ω, what should be the value of R_L for maximum power transfer.

(A) 3Ω (B) 5Ω

(C) Zero (D) None of these.

Q.13. In the network of Fig. 13.4, when will be the power transferred across terminals (X-Y) maximum ?

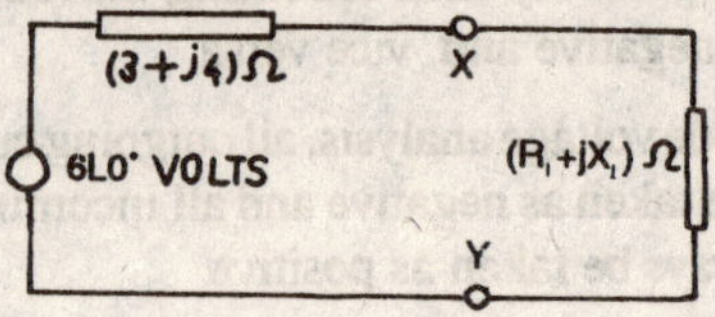

Fig. 13.4.

(A) $R_L = 3\Omega$ and $X_L = -4\Omega$

(B) $(R_L+jX_L)=3+j4$

(C) $R_L = 5\Omega$

(D) None of these.

Q.14. Fig. 13.5 shows Norton's equivalent of a linear active network across the terminals of interest. If we had drawn the Thevenin's equivalent of the same network across same terminals, what would have been the magnitude of Thevenin's generator ?

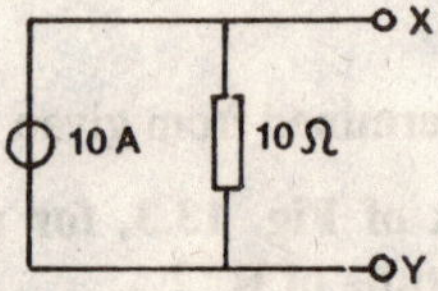

Fig. 13.5.

(A) 100 volts

(B) j 100 volts

(C) 1 volt

(D) It cannot be determined from given information.

Q. 15. From the network of Fig. 13.6 find out its Norton's equivalent across X-Y. What should be the Norton's current magnitude ?

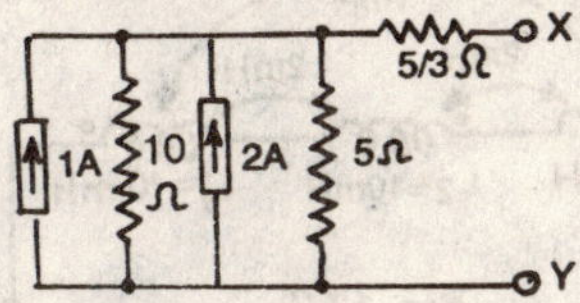

Fig. 13.6.

(A) 1 A (B) 3 A

(C) 2 A (D) Cannot be determined.

Q. 16. The FINAL VALUE theorem with reference to Laplace transform method of Network Analysis is

(A) Expressed by $f(0^+) = \lim_{S \to 0} [SF(s)]$

(B) Expressed by $f(\infty) = \lim_{S \to \infty} [SF(s)]$

(C) Expressed by $f(\infty) = \lim_{S \to 0} [F(s)]$

(D) Expressed by $f(\infty) = \lim_{S \to 0} [SF(s)]$.

Q. 17. What is equivalent inductive reactance between X-Y in the arrangement of Fig. 13.7.

(A) j6Ω

(B) j9Ω

(C) j12Ω

(D) j10Ω.

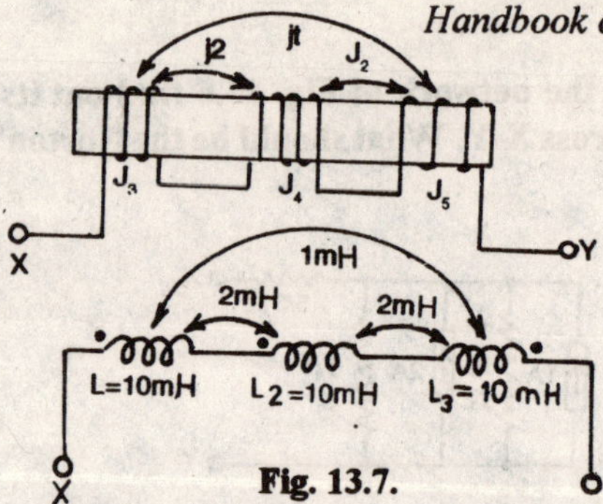

Fig. 13.7.

Q. 18. From the coil arrangement of Fig. 13.7. What in your opinion is the effective inductance between terminals X-Y?

(A) 30 mH (B) 0 mH

(C) 28 mH (D) 29 mH.

Q. 19. In the network of Fig. 13.8 $R_1 = R_2 = 2000\Omega$ and $R_3 = 10\Omega$. What in your opinion should be (n) to obtain maximum output voltage (V_0) ?

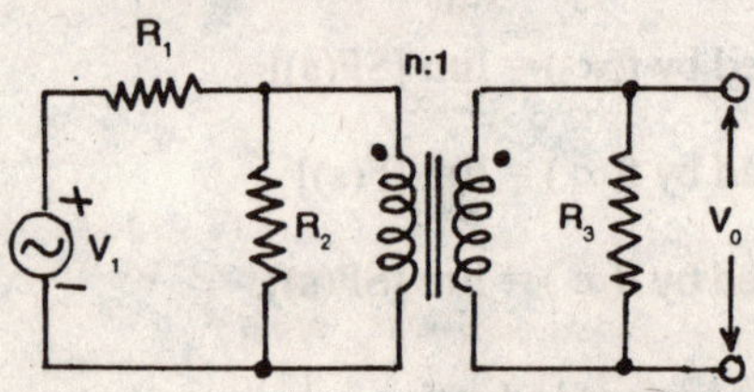

Fig. 13.8.

(A) 10 (B) 100

(C) 1 (D) None of these.

Q. 20. What is (Z_1) in case of the network of Fig. 13.9

(A) 12Ω (B) 6Ω

(C) 2Ω (D) 1Ω

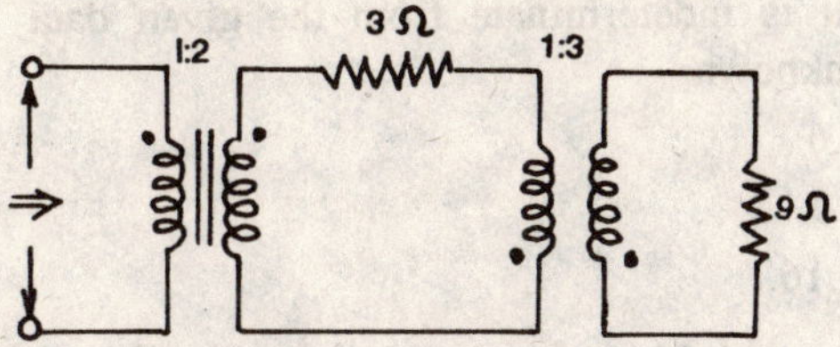

Fig. 13.9.

Q. 21. From the R-L-C network of Fig. 13.10 Where will you place the dot for coil L_1 (at A or B) so as to make it series resonant?

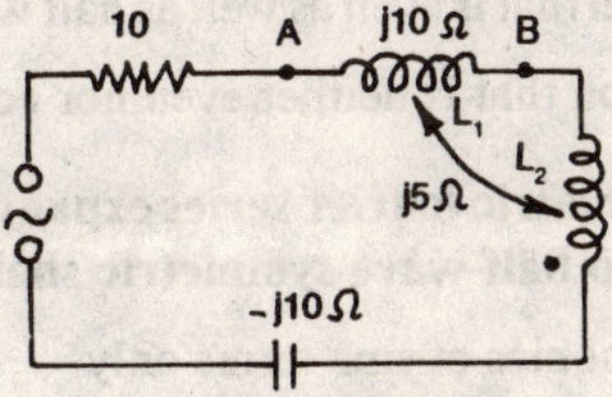

Fig. 13.10.

(A) At A (B) At B

(C) None of these.

Q. 22. What is the co-efficient of coupling (K) between the two coils in the arrangement of Fig. 13.11.

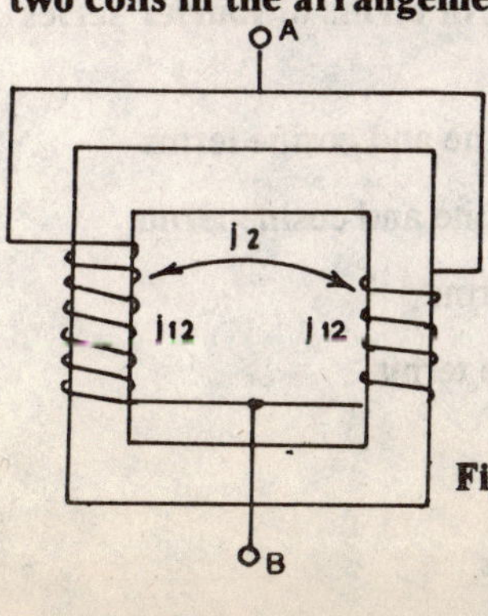

Fig. 13.11.

(A) It is indeterminate from the given data ω is unknown

(B) 0.6

(C) 0.1

(D) 0.16.

Q. 23. A sinusoidal waveform is mathematically represented by

(A) An EVEN function

(B) An ODD function

(C) A function that is even as well as half wave symmetric

(D) A function that is neither even nor odd.

Q. 24. The trigonometric fourier series expansion of an even function that is also half-wave symmetric shall contain

(A) Odd harmonics of sine terms only

(B) Both sine and cosine terms

(C) Only cosine terms

(D) Only odd harmonics of cosine terms.

Q. 25. A function that is neither even nor odd but has half-wave symmetry can be expressed in terms of fourier series expansion as a sum of

(A) Odd harmonics of both sine and cosine terms

(B) Even harmonics of both sine and cosine terms

(C) Odd harmonics of sine terms

(D) Even harmonics of cosine terms.

ANSWERS

1. B,	2. B,	3. B,	4. A,	5. A,	6. B,
7. A,	8. C,	9. B,	10. B,	11. D,	12. A,
13. A,	14. A,	15. C,	16. C,	17. A,	18. C,
19. A,	20. D,	21. A,	22. D,	23. B,	24. D,
25. A.					

Chapter 14

Linear Integrated Circuits Applications

STUDY MATERIAL

An Operational Amplifier

It is a high gain differential amplifier having a differential input and normally a single ended output.

An Ideal Opamp

It shall have an infinite open loop voltage gain (Even practical opamps have open loop gain as high as 100 dB or more).

If one of the inputs of the opamp is grounded, the other cannot deviate from it. Feedback is introduced in opamps to have desired and controlled specifications. Introduction of feedback in opamps has made possible its applications as amplifiers, integrators, differentiators, function generators, multivibrators, oscillators, power supply regulators, gyrators, active filters and so on.

Introduction of negative feedback reduces the gain, increases bandwidth, increases input impedance and reduces output impedance.

Slew Rate

It is one of the most important characteristics of opamps is the rate of change of output voltage with time. It is normally expressed in volts per microsecond (0.5 V/μs in case of 741). Higher the slew rate, better is the opamp.

A timer

It is a digital IC in which two opamp comparators control the operation of a Set/Reset flip flop.

Timer in various circuit configurations can be used as (i) Frequency Dividers (ii) Pulse width modulators (iii) Pulse position modulators (iv) Tone burst generator (v) Temperature control where the temperature change appears as a change in output frequency, (vi) d.c. to d.c. converter (vii) Ramp generator (viii) Audio oscillator (ix) Missing pulse detector (x) Transformerless negative voltage generator and many many more.

Voltage Regulators

Both linear as well as switching, are completely dominated by linear ICs.

Basic constituents of a Power Supply Regulator

It has four basic constituents namely a reference source, an error detector, a control device and protection circuitry.

The free running frequency or the centre frequency of a PLL

It is the frequency at which the PLL VCO operates when it is not locked to the input.

Phase-comparator gain

It relates the phase comparator output voltage to the phase difference between the VCO output and input signal when the loop is in lock. It is expressed in volts/radian. VCO conversion gain relates oscillator frequency shift from the center frequency to the applied input voltage. It is expressed in Radians per second per volt.

OBJECTIVE TYPE QUESTIONS

Q. 1. Which one of the following configurations is that of a true voltage follower?

(A) The one given in Fig. 14.1 (d) is correct

(B) The configurations of Fig. 14.1 (a) and (b) are correct.

(C) The one given in Fig. 14.1 (a) is correct

(D) The one given in Fig. 14.1 (b) is correct

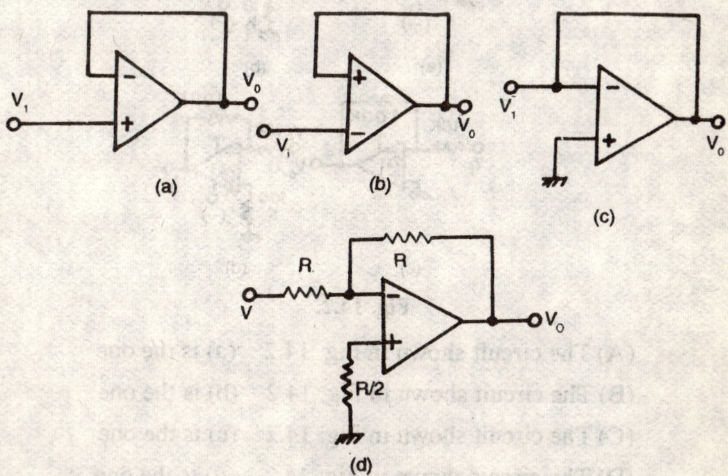

Fig. 14.1.

Q. 2. Can we say that the input impedance of an opamp voltage follower would always decrease with increase in frequency?

(A) Yes, we can say

(B) No, it is not true. The input impedance, infact, increases with increase in frequency

(C) Input impedance of a voltage-follower is independent of frequency

(D) None of these.

Q. 3. Figs. 14.2 (a), (b), (c) and (d) shows three different inverting amplifier configurations. Each one of them provides a gain of 10. Select the one that is the most correct configuration.

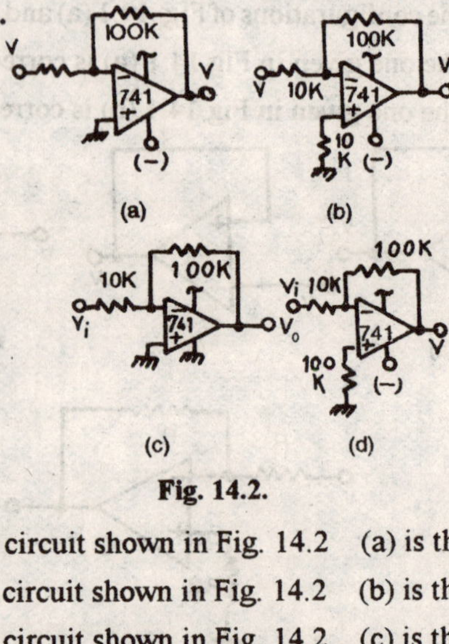

Fig. 14.2.

(A) The circuit shown in Fig. 14.2 (a) is the one

(B) The circuit shown in Fig. 14.2 (b) is the one

(C) The circuit shown in Fig. 14.2 (c) is the one

(D) The circuit shown in Fig. 14.2 (d) is the one.

Q.4. What do we mean by an internally and fully compensated opamp?

(A) This means that the opamp has internal circuit components which ensure that the opamp is always stable (even in case of 100% feedback)

(B) This means that the opamp's frequency response has already been extended to the maximum internally

(C) This means there is no need for connecting external compensation network

(D) This means that the IC's internal circuitry contains the compensation capacitance.

Q. 5. What will be the output in case of the integrator shown in Fig. 14.3?

(A) $\int_0^i v_i \, dt$ (B) $-\int_0^t v_i \, dt$

(C) $\frac{1}{RC}\int_0^t v_i \, dt$ (D) $-\frac{1}{RC}\int_0^t v_i \, dt$

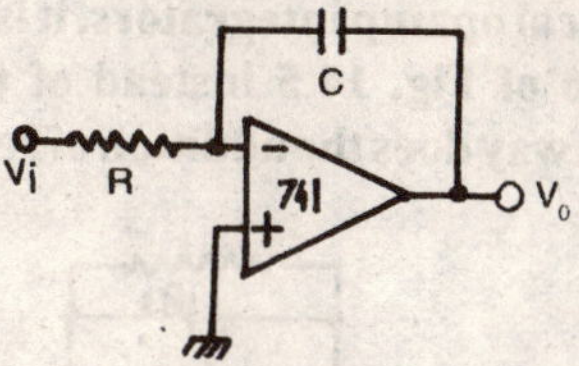

Fig. 14.3.

Q. 6. In an opamp inverting amplifier configuration with an input resistance (R_1) and a feedback resistance (R_2), it is a normal practice to connect a resistance (R) equal to parallel combination or (R_1) and (R_2) from the non-inverting input to ground. This is done to

(A) Compensate for the effects of input offset voltage

(B) Compensate for the effects of input offset current

(C) Compensate for the effects of input bias currents

(D) Improve upon CMRR rating of opamp.

Q.7. Fig. 14.4 shows an opamp voltage follower circuit. This circuit is being driven by another circuit with output resistance of 100 kΩ. What should be the value of R?

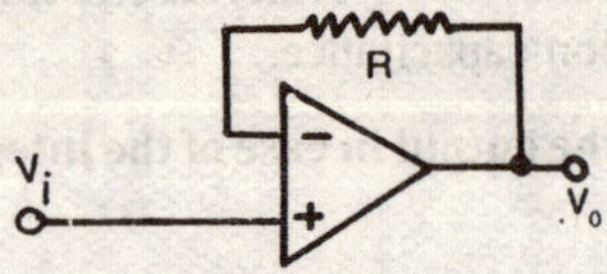

Fig. 14.4.

(A) 100 kΩ (B) 0

(C) 10 kΩ (D) 1 MΩ.

Q.8. In all practical opamp integrators, it is preferable to use the configuration of Fig. 14.5 instead of the one shown in Fig. 14.3. In what way does the addition of resistance (R_3) help

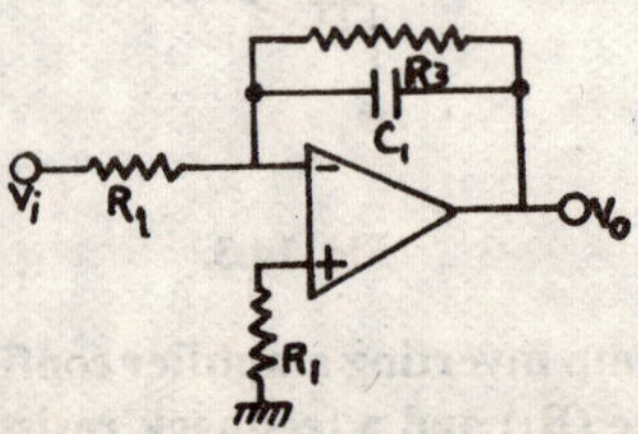

Fig. 14.5.

(A) It helps in linearizing the output waveform incase input is a square

(B) It improves upon d.c. stability by providing a finite d.c. gain

(C) It overcomes bad effects of input offset currents and voltages

(D) It reduces transients.

Q. 9. In a conventional astable multi-vibrator using 555 [Refer to Fig. 14.6], the frequency of the output waveform will be given by

(A) $1.45/(R_1 + 2R_2)\,C$ (B) $1/(R_1 + 2R_2)\,C$

(C) $\dfrac{1.45}{(R_1 + 2R_2)\,C}$ (D) $\dfrac{1}{1.1\,(R_1 + 2R_2)\,C}$

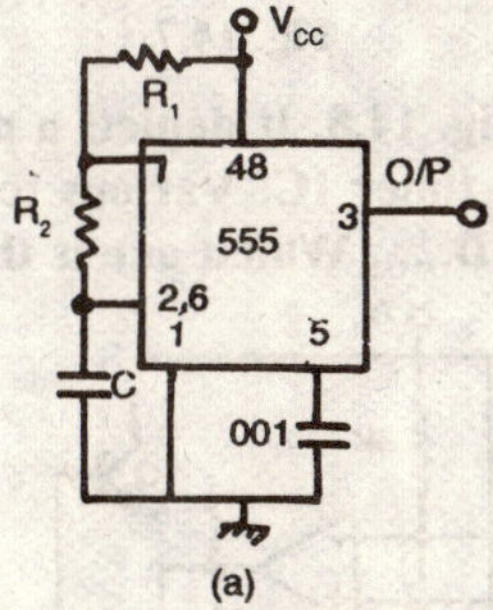

Fig. 14.6.

Q. 10. The configuration of Fig. 14.7 represents

(A) It is an astable multivibrator, with R_2 and C being timing components

(B) It is a monoshot with pulse width given by $1.1\,R_1\,C$

(C) It is an astable circuit with charging taking place through R_1 and discharge taking place through R_2.

(D) It is an astable circuit in which charging takes place when the output is low.

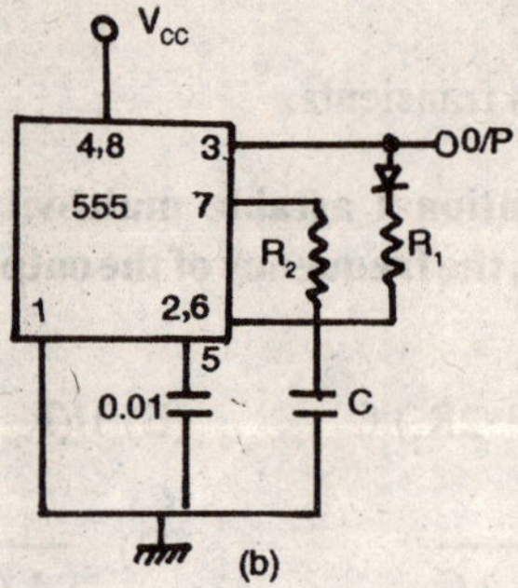

Fig. 14.7.

Q. 11. Refer to Fig. 14.8. It depicts a part of the internal circuitry of a 555 timer IC. Various terminals have been labelled as A,B,C,D...... Which one is the control terminal (pin 5 of 555)?

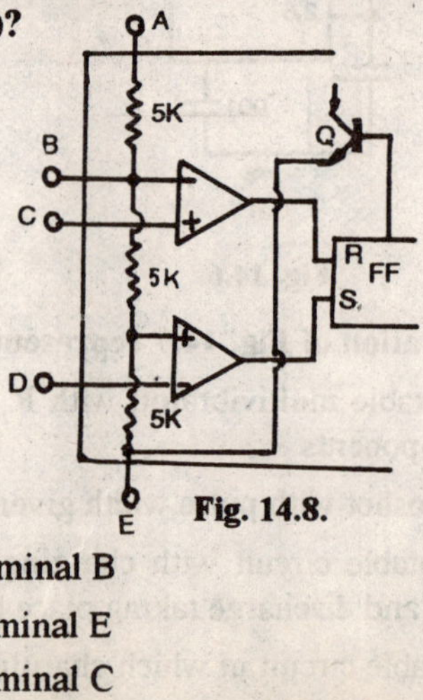

Fig. 14.8.

(A) Terminal B

(B) Terminal E

(C) Terminal C

(D) Control terminal has not been included in this diagram.

Q. 12. Refer to Fig. 14.9. Assume diode D to be ideal. In this free running multi-vibrator configuration, the timing capacitor (C) charges and discharges during operation between

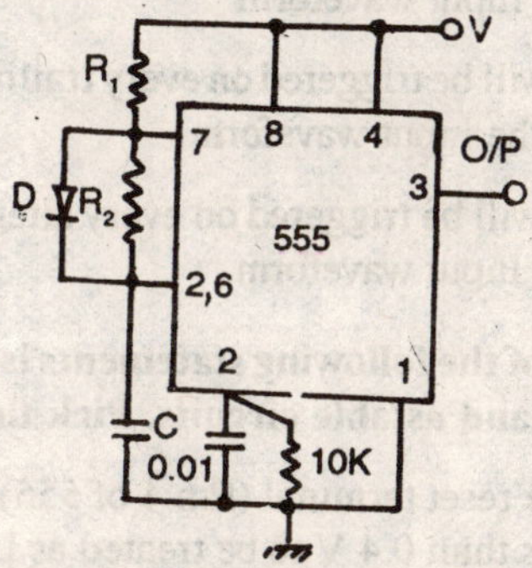

Fig. 14.9.

(A) 0 and $V_{CC}/3$ (B) $V_{CC}/3$ and Vcc

(C) 1/3 V_{CC} and 2/3 V_{CC} (D) 1/4 V_{CC} and 1/2 V_{CC}.

Q. 13. In the kind of input triggering arrangement shown for the monoshot of Fig. 14.10, at which time instants will the triggering take place assuming that the output pulse width has been chosen as per R and C values to be equal to 50 μS?

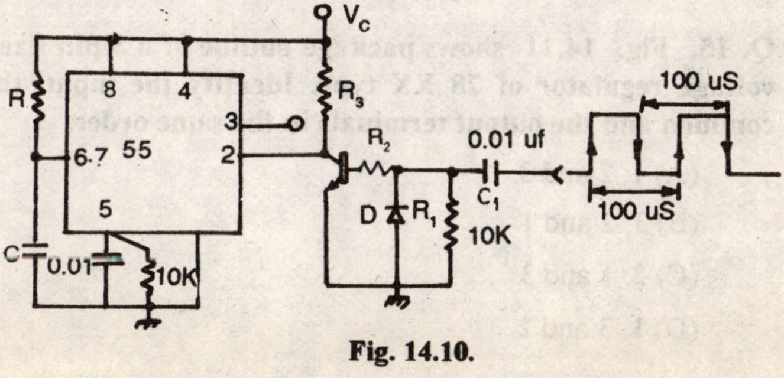

Fig. 14.10.

(A) It will be triggered on every rising edge (low to high) of input waveform

(B) It will be triggered on every alternate leading edge of the input waveform

(C) It will be triggered on every trailing edge (high to low) of the input waveform.

(D) It will be triggered on every alternate trailing edge of the input waveform.

Q. 14. One of the following statements is false regarding 555 monostable and astable circuits. Pick that out

(A) The reset terminal (Pin 4 of 555) must have a voltage less than 0.4 V to be treated as LOW

(B) In a 555 monoshot, the input trigger pulse width should always be less than the desired output pulse width.

(C) A 555 monoshot can be made to trigger either on the leading edge or on the trailing edge of the input waveform

(D) The output waveform in a 555 monoshot has a frequency equal to that of the input trigger waveform but with a different duty cycle.

Q. 15. Fig. 14.11 shows package outline of a 3-pin fixed voltage regulator of 78 XX type. Identify the input, the common and the output terminals in the same order.

(A) 1, 2 and 3

(B) 3, 2 and 1

(C) 2, 1 and 3

(D) 1, 3 and 2.

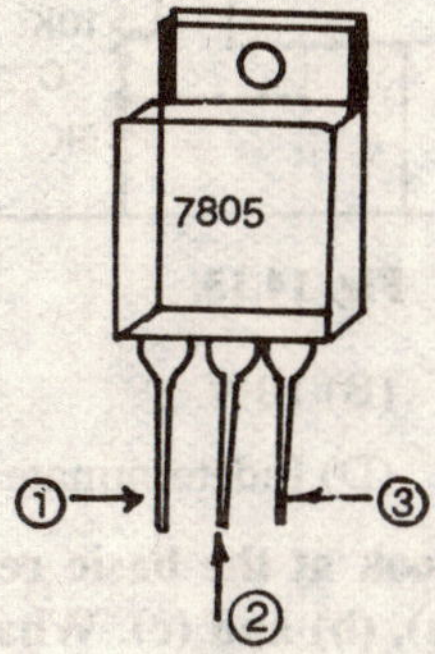

Fig. 14.11.

Q. 16. Fig. 14.12 shows the arrangement in which a 3-pin positive regulator has been used as a constant current regulator. What will be this constant output current in this case?

(A) $I_0 \simeq 550$ mA (B) $I_0 \simeq 1$ A

(C) $I_0 \simeq 550$ mA (D) I_0 is indeterminate.

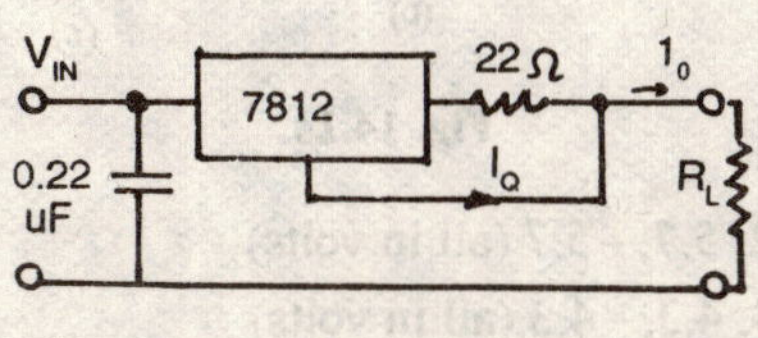

Fig. 14.12.

Q. 17. What is the output voltage (V_0) in the circuit of Fig. 14.13. The control terminal voltage is 5 V.

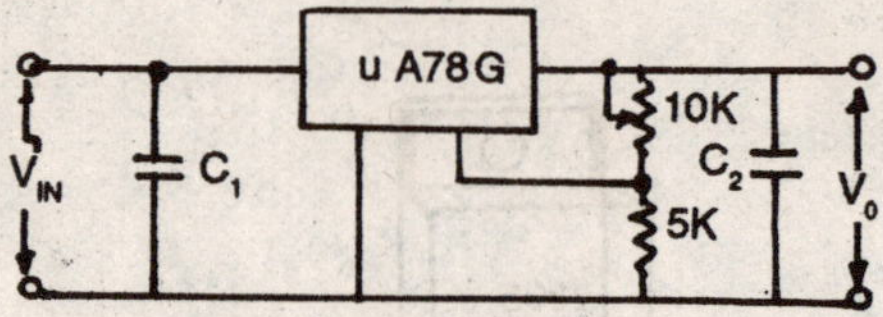

Fig. 14.13.

(A) 5 V (B) 10 V

(C) 15 V (D) Indeterminate from given data.

Q. 18. Have a close look at the basic regulator circuits, shown in Fig. 14.14 (a), (b) and (c). What are the output voltages. Your first answer should be for the circuit of Fig. 14.14 (a) and then for the circuit of Fig. 14.14 (b) and finally for that of 14.14 (c).

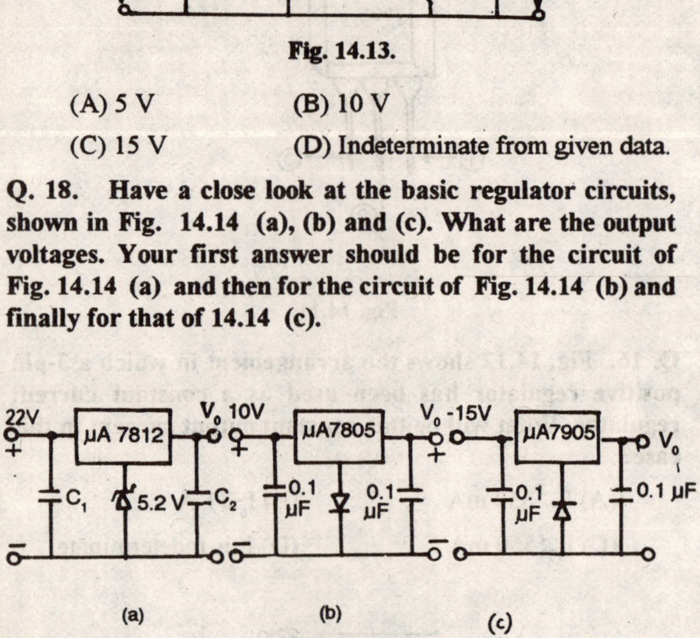

Fig. 14.14.

(A) 17.2, 5.7, - 5.7 (all in volts)

(B) 17.6, 4.3, - 4.3 (all in volts)

(C) 17.6, 4.3 - 5.7 (all in volts)

(D) 6.4, 5.7, - 5.7 (all in volts).

Q. 19. When a PLL is being used as phase modulator, the modulating signal is injected

(A) At the input of PLF

(B) At the input of phase comparator

(C) At the input of error amplifier

(D) From the output of LPF.

Q. 20. In PLL 562 application as a frequency translator, a Divide by 10 counter is externally inserted between VCO output and phase comparator input. If the input is 100 KHz, the VCO output will be

(A) A 100 KHz signal (B) A 10 KHz signal

(C) A 1 MHz signal (D) A 10 MHz signal.

ANSWERS

1. C,	2. A,	3. B,	4. A,	5. D,	6. C,
7. A,	8. B,	9. A,	10. A,	11. A,	12. D,
13. A,	14. D,	15. A,	16. A,	17. C,	18. A,
19. D,	20. C.				

Chapter 15

Pulse Techniques

STUDY MATERIAL

A high pass RC circuit

Such a circuit allows only those frequencies that are above the cut-off value given by $f_c = \frac{1}{2\pi RC}$. It behaves as a differentiator if the time constant of the circuit is much smaller than the period of the applied input.

A *pulse* when applied to a differentiator circuit produces two spikes: a positive spike at the leading edge and a negative spike at the trailing edge.

A *sinusoidal input* to a high pass circuit gives an output whose amplitude is a function of frequency given by the equation

$$|A| = \frac{1}{\sqrt{1+\left(\frac{f}{f_H}\right)^2}}$$

where $f_H = \frac{1}{2\pi .\ RC}$

and $|A|$ = Magnitude of gain.

A low pass RC circuit

It allows only those frequencies that are below a certain cut-off frequency given by $fc = \frac{1}{2\pi RC}$. It behaves as an integrator if the time constant (RC) is much larger than the time period of the applied input.

A sinusoidal input to a low pass circuit gives an output whose amplitude is a function of frequency given by the equation

$$|A| = \frac{1}{\sqrt{1+\left(\frac{f_L}{f}\right)^2}}$$

where $f_L = \frac{1}{2\pi RC}$

and $|A|$ = Magnitude of gain.

In a *bistable multivibrator*, both the states (ON as well as OFF) are stable. An *R-S flip-flop* is basically a bistable multivibrator. *Schmitt trigger* circuit is also a modified version of a bistable multivibrator.

In an *astable multivibrator*, none of the states is stable. The circuit shifts from one state to another and functions in a free running mode. In a *symmetrical astable multivibrator*, the off-period equals the on-period. Each is given by (0.69 RC) Mathematically, the frequency, of output is given by

$$f = \left(\frac{1}{1.38\ RC}\right).$$

A *pulse-transformer.* It is a broad band transformer. It is supposed to preserve the shape of the input pulse. For this, the core material should be the one of very high permeability and very high resistivity. It should have the least leakage inductance and shunt capacitance. Leakage inductance and shunt capacitance determine the rise time response of the pulse-transformer. Mathematically, rise time is given by the following equation.

$$t_r \propto 2\pi (a\sigma c)^{1/2} = k \times 2\pi(a\sigma c)^{1/2}$$

where $\rho \rightarrow$ Leakage Inductance

$c \rightarrow$ Shunt capacitance

$a \rightarrow$ Damping factor

Also $a = \dfrac{R_1}{R_r + R_2} \approx$ as $R_2 \gg R_1$

where R_1 = Sum of source resistance and primary winding resistance.

R_2 = Secondary resistance plus load resistance reflected to the primary.

The constant of proportionality in the above equation depends upon $R\left(= \dfrac{R_1 R_2}{R_1 + R_2}\right)$, c, σ, etc. In a critically damped case, it is unity. For an underdamped case it is less than unit and it exceeds unity for an overdamped case.

The flat-top response and the backswing in a pulse transformer

These are determined by the magnetizing inductance or more precisely the ratio of (Rtp) to (L). Mathematically,

$$\text{Percentage tilt} = \left(\frac{Rtp}{L}\right) \times 100$$

where $R = \left(\dfrac{R_1 R_2}{R_1 + R_2}\right)$

Leakage inductance (σ)

It can be determined by short circuiting the secondary and measuring the inductance across input terminals. It can also be determined by the leakage inductance to shunt the primary winding with a capacitor (with secondary shorted) and measure the resonant frequency by forming it as a part of collector load of a transistor circuit and switching the transistor ON and OFF.

Another reading is then taken with a different value of capacitor. If (f_1) and (f_2) are the resonant frequencies with (c_1) and (c_2), then the leakage inductance (σ) is given by

$$\sigma = \left[\frac{f_1^2 - f_2^2,}{(2\pi f_1 f_2)^2 \, (c_2 - c_1)} \right]$$

The average value of a train of pulses of frequency (f) and peak value V_p is given by

$$V_{av} = V_p \left(\frac{T_{ON}}{T_{ON} + T_{OFF}} \right)$$

where $T_{ON} + T_{OFF} = T = \left(\frac{1}{f} \right)$

$$\text{Also duty cycle} = \frac{T_{ON}}{T_{ON} + T_{OFF}}$$

IC 555

It is a timer IC that can be used in various applications. Basically, it operates in either of the two modes, (i) Monostable mode (ii) Astable mode. Internally, it consists of two comparators, two transistors, three identical resistors and an output stage. It can operate at all supply voltages from + 5 V to +18 V making it compatible with both opamps and TTL circuits. A single 555 has a maximum timing range of about 15 minutes but it can be extended by cascading. It is available in following types of packages(i) TO99 (ii) DIP. It has 8-pins with

PIN 1→ Ground.

PIN 2→ Trigger terminal.

PIN 3→ Output terminal

PIN 4→ Reset terminal.

PIN 5→ Control terminal.

PIN 6→ Threshold point.

PIN 7→ Discharge terminal.

PIN 8→ Positive V_{cc} (5 V to 18 V)

The output terminal (PIN 3) of 555 presents a resistance of about 10Ω between PIN 3 and ground when the IC is in the low state and a resistance of about the same value between pin 3 and positive V_{cc} (pin 8) when the IC is in the high state. Normally, the IC can supply or sink a current of 10 mA. Maximum current which the IC can source or sink is 200 mA. Also, in the high state, the output is slightly less then + V_{cc} (by 0.5 V) and in the low state, it is 0.1V.

The trigger terminal (pin 2) if kept low keeps the output high and if kept at about $\frac{2}{3}V_{cc}$ keeps it low. A negative trigger of sufficient amplitude brings the output to the high state.

The reset terminal (pin 4) is joined with + V_{cc} if it is not used. V_{cc} at pin 4 keeps the output high. A low voltage at pin 4 brings both the output terminal (pin 3) and discharge terminal (pin 7) to ground potential.

The control terminal (pin 5) can be used to charge the threshold and trigger levels. A.0 01 μF capacitor is generally connected from pin 5 to ground. This capacitor bypasses any high frequency noise.

When IC 555 is used in the Astable made, the frequency of the output is given by

$$f = \frac{1.44}{(R_1 + 2R_2)c}$$

whese R_1 is resistance connected between pin 7 and pin 8, R_2 is resistance connected between pin 7 and the junction of 2 and 6, C is the capacitor from pin 6 to ground. It may also be mentioned that the capacitor charges through R_1 and R_2 whereas it discharges through R_2 only.

The duty cycle is given by $\frac{R_2}{R_1 + R_2}$ and a duty cycle of 50% means $R_1 = 0$ which is not feasible as it would short pin 8 and pin 7 resulting in a heavy current when the discharge transistor is ON. Also, a duty cycle of more than 50% is impossible to achieve with this type of circuit. Duty cycles of $\gg$ 50% can be achieved by connecting a diode across R_2. Duty cycle in this particular case is $\left(\frac{R_2}{R_1 + R_2} \right)$. Here duty cycle has been defined as $\frac{T_{low}}{T_{low} + T_{high}}$.

Timer IC 555 can also be used as a Pulse-Width-Modulator (PWM) when connected in the monostable mode. 555 is triggered by continuous train of pulses and the control voltage is applied at the control terminal (pin 5). The control signal varies the threshold level and hence the width of the output pulse.

Timer IC 555 can also be used as a pulse position modulator (PPM) when connected in the astable mode. Control voltage is applied at pin 5 again. Here also, the control voltage varies the threshold level.

OBJECTIVE TYPE QUESTIONS

Q. 1. Figure 15.1 shows an input applied to a circuit, whether it is a differentiator or an Integrator ?

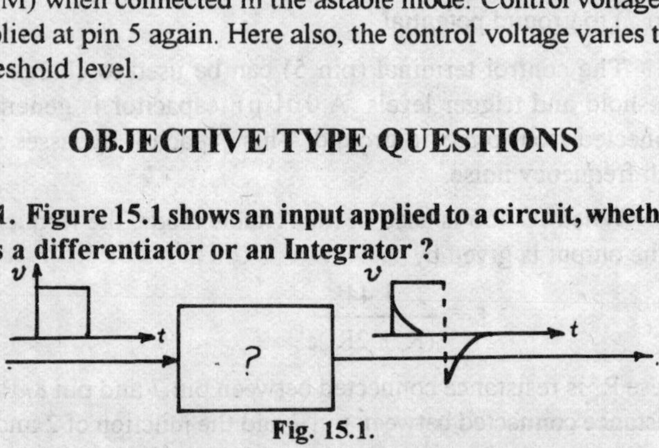

Fig. 15.1.

(A) Differentiator (B) Integrator

(C) High pass (D) Low pass.

Q. 2. Figure 15.2 shows an input applied to an unknown R-C circuit to yield an output. What is the nature of the circuit?

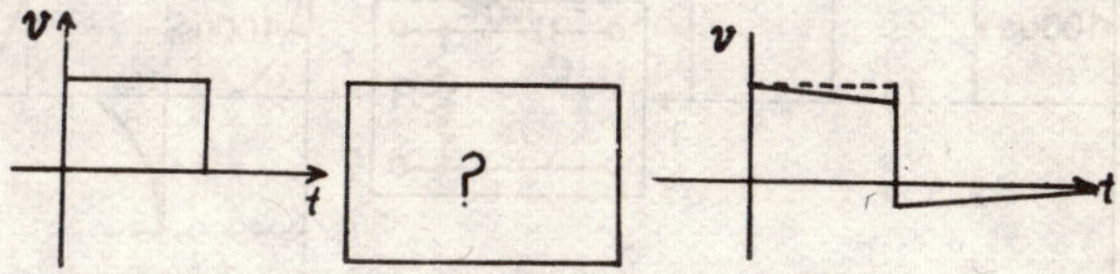

Fig. 15.2.

(A) It is a high-pass circuit with a very small time constant (i.e., RC $< t_p$)

(B) It is a high pass circuit with a very large time constant (RC $> t_p$)

(C) It is an R-C circuit with RC $\approx t_p$

(D) None of these.

Q. 3. Figure 15.3 shows an input, an output and a circuit block. What is the value of C?

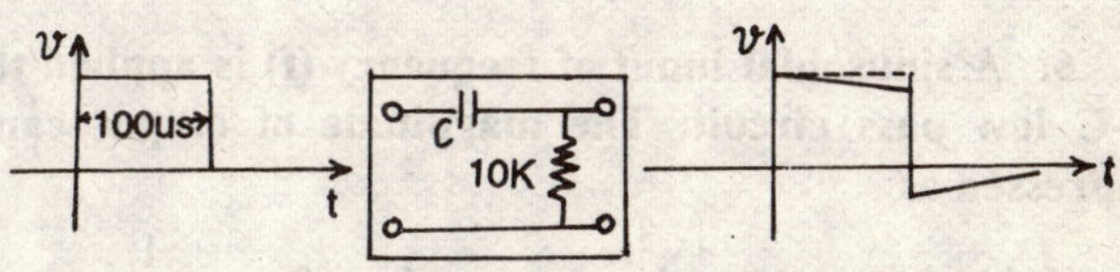

Fig. 15.3.

(A) 1 μF (B) 1000 μF

(C) 4 μF (D) 50 pF.

Q. 4. Figure 15.4 shows an input, an output and a circuit block. What is the value of R?

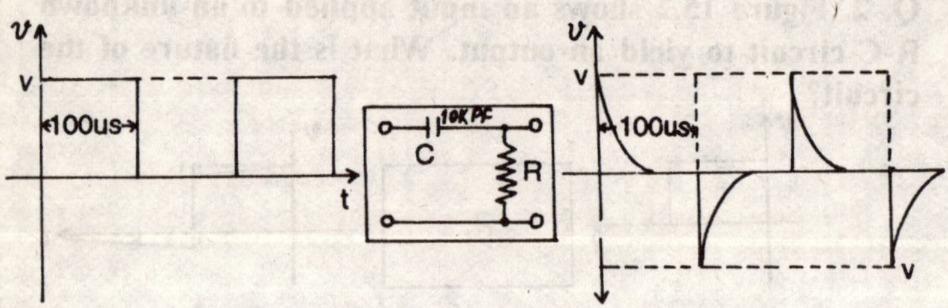

Fig. 15.4.

(A) 150Ω (B) 100 kΩ

(C) 5 MΩ (D) 450 kΩ.

Q. 5. When a low-pass circuit behaves as an integrator, the R-C product is such that

(A) It is much larger than the time period of input wave form

(B) It is equal to the time period of the input waveform

(C) It is much smaller than the time period of the input waveform

(D) None of these.

Q. 6. A sinusoidal input of frequency (f) is applied to an R-C low pass circuit. The magnitude of output can be expressed as

(A) $|v_0| = \dfrac{|v_i|}{\sqrt{1+(f/f_2)^2}}$ where $f_2 = \dfrac{1}{2\pi RC}$

(B) $|v_0| = \dfrac{v_i}{\sqrt{1+(f_2/f)^2}}$ where $f_2 = \dfrac{1}{2\pi RC}$

(C) $|v_0| = \dfrac{v_i}{\sqrt{1-(f/f_2)^2}}$ where $f_2 = \dfrac{1}{2\pi RC}$

(D) None of these.

Q. 7. Figure 15.5 shows an input, an output and a circuit block. What is the value of C ?

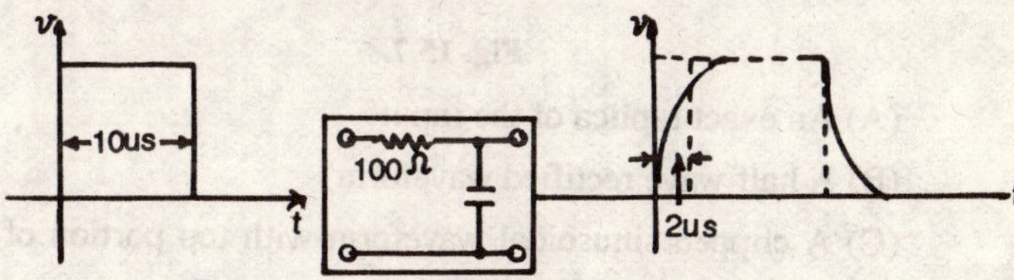

Fig. 15.5.

(A) 10 KPF (B) 5 μF

(C) 10 μF (D) 50 pF.

Q. 8. Figure 15.6 shows an input, an output and a circuit block. What is the value of R?

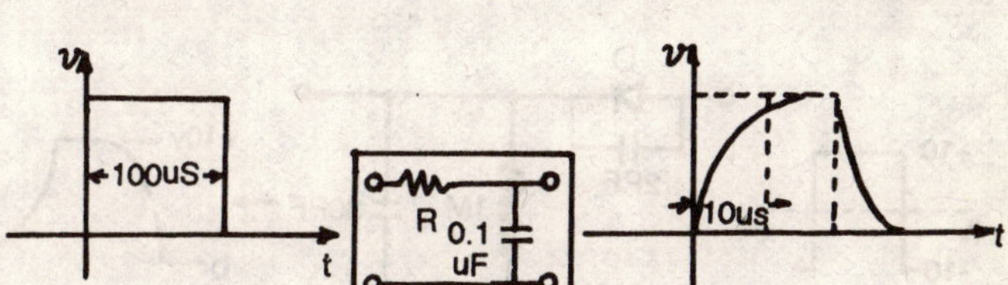

Fig. 15.6.

(A) 4 K (B) 4 M

(C) 470 K (D) 150 Ω.

Q. 9. Figure 15.7 shows a diode clipper circuit. A sinusoidal waveform is applied at its input. The output will be given by

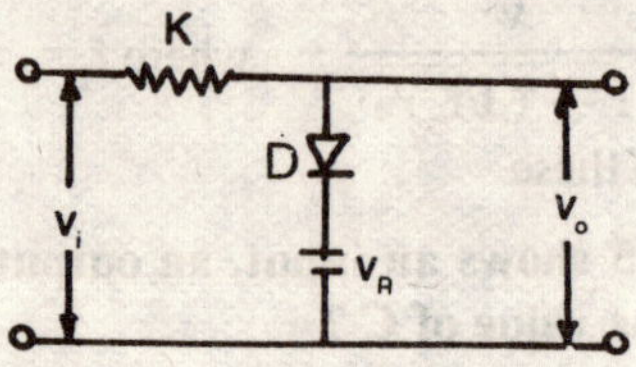

Fig. 15.7.

(A) An exact replica of the input

(B) A half wave rectified waveform

(C) A clipped sinusoidal waveform with top portion of positive half cycles clipped

(D) A clipped sinusoidal waveform with lower portion of negative half cycles removed.

Q. 10. Figure 15.8 shows a diode clipper circuit. The input is a pulse as shown. The clipped output has also been shown. The rise time of the output pulse will be (Assume on resistance of diode to be 100)

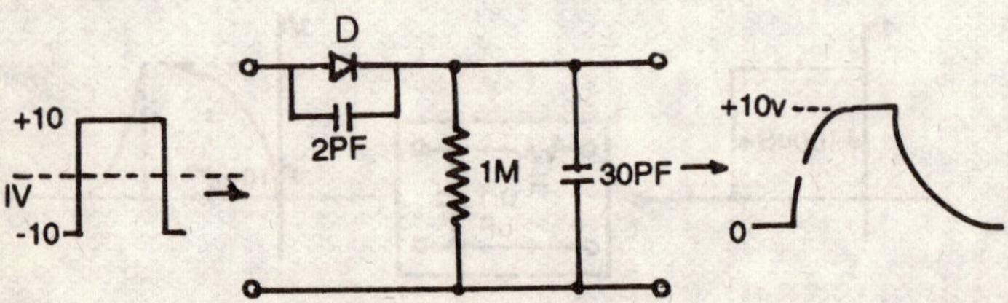

Fig. 15.8.

(A) 3 ~ ns (B) ~ 6 ns

(C) ~ 4 μs (D) 10 μs.

Q. 11. The fall time of the output pulse in the question 10 is

(A) ~30 μs (B) ~100 μs

(C) ~1 μs (D) ~40 ns.

Q. 12. Circuit shown in Fig 15.9 is a basic

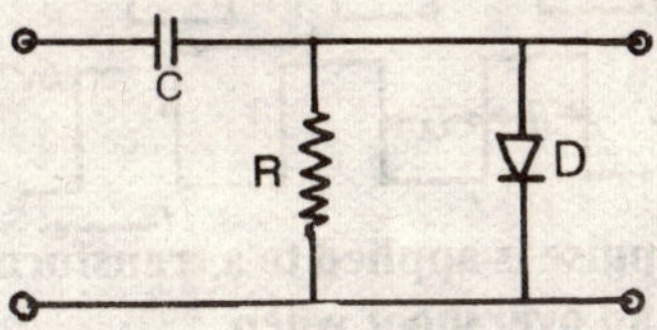

Fig. 15.9.

(A) Clipping circuit (B) Clamping circuit

(C) Two level clipper (D) None of these.

Q. 13. What is the output waveform in the circuit shown in Fig. 15.10 [Assume V_{cc} (sat) = OV]

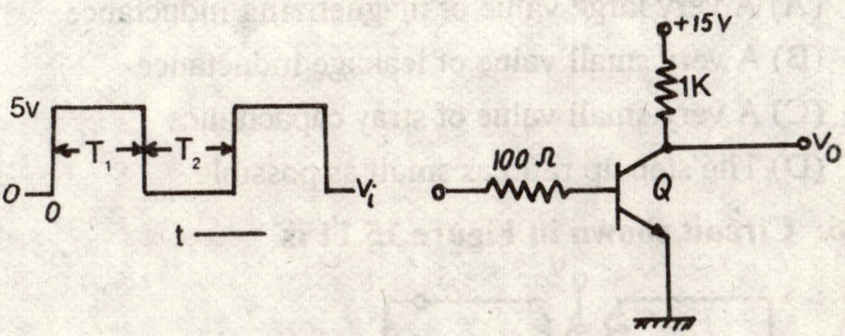

Fig. 15.10.

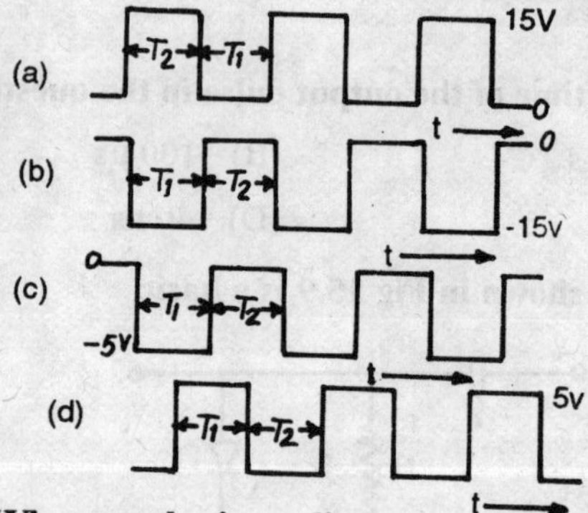

Q. 14. When a pulse is applied to a transformer, the output pulse will have an overshoot when

(A) The pulse transformer is overdamped

(B) The pulse transformer is underdamped

(C) The pulse transformer is critically damped

(D) None of these.

Q. 15. In order to reduce the droop in the flat top portion of the pulse and the backswing at the end of the pulse, it is required to have

(A) A very large value of magnetizing inductance

(B) A very small value of leakage inductance

(C) A very small value of stray capacitance

(D) The step up ratio as small as possible.

Q. 16. Circuit shown in Figure 15.11 is

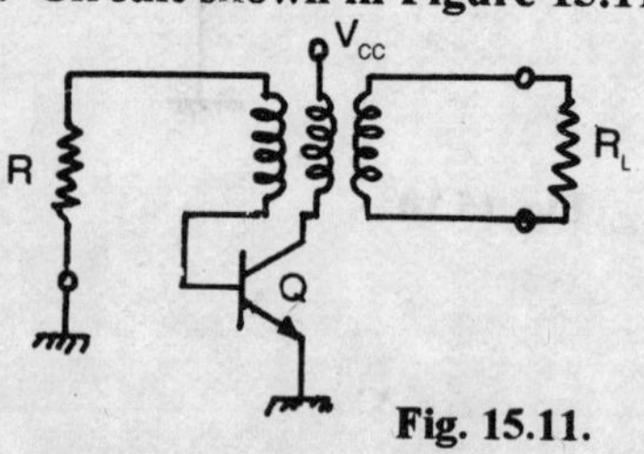

Fig. 15.11.

(A) A monostable multi-vibrator

(B) A current sweep generator

(C) A tuned amplifier

(D) A monostable blocking oscillator.

Q. 17. Circuit shown in Figure 15.12 is

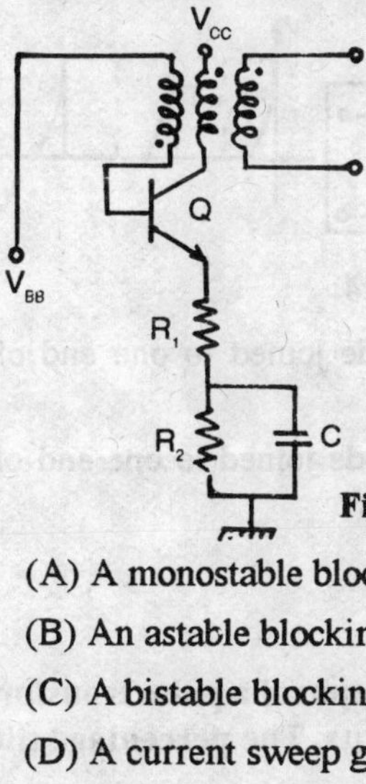

Fig. 15.12.

(A) A monostable blocking oscillator

(B) An astable blocking oscillator

(C) A bistable blocking oscillator

(D) A current sweep generator.

Q. 18. Figure 15.13 shows a pulse train. The average voltage of this pulse train is equal to

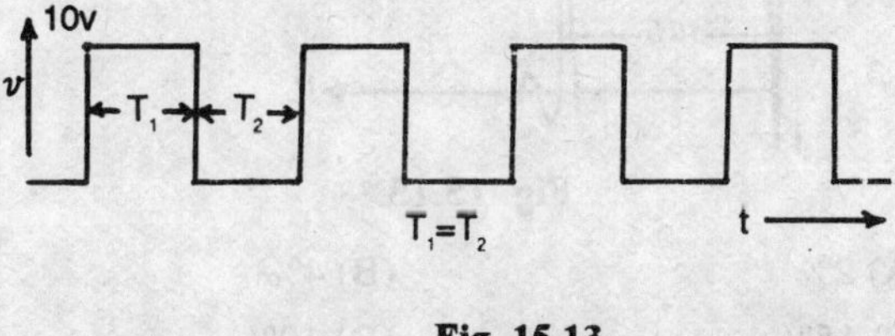

Fig. 15.13.

(A) 40 V (B) 8.66 V

(C) 5 V (D) 3.33 V.

Q. 19. Figure 15.14 shows an input, an output and a circuit block. One of the circuit elements is missing. What is that?

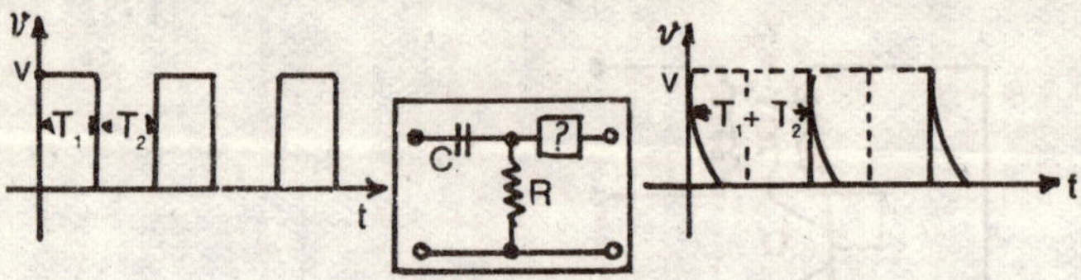

Fig. 15.14.

(A) It is a diode with anode joined to one end of the resistor R

(B) It is a diode with cathode joined to one end of the resistor R

(C) It is a capacitor

(D) It is a resistor.

Q. 20. Figure 15.15 shows the output of a pulse transformer with L = 5 mH, R = 100Ω, t_p = 1 μs. The percentage tilt is

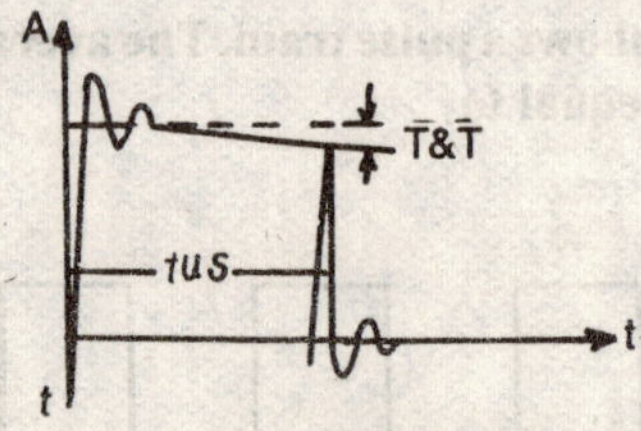

Fig. 15.15.

(A) 2% (B) 4%

(C) 15% (D) 10%.

Q. 21. IC 555 is

(A) An operational amplifier

(B) A timer IC

(C) A single shot

(D) None of these.

Q. 22. Figure 15.16 shows the application of 555 timer circuit. What circuit it is?

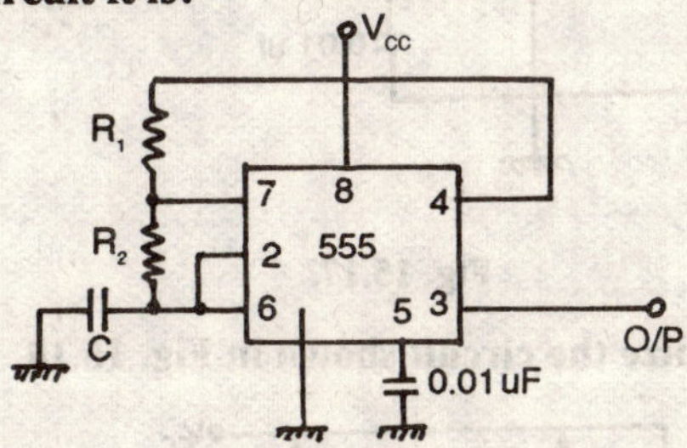

Fig. 15.16.

(A) Monostable multi-vibrator

(B) Astable multi-vibrator

(C) Schmitt trigger

(D) None of these.

Q. 23. Figure 15.17 shows an astable mode of operation of 555. What should be the condition so that the circuit generates a square wave.

(A) $R_1 = R_2$ (B) $R_1 = 2R_2$

(C) $R_1 = R_2 = 10$ K, $C = 1\ \mu F$ (D) $R_1 = 10$ K, $R_2 = 20$ K, $C = 0.1\ \mu F$

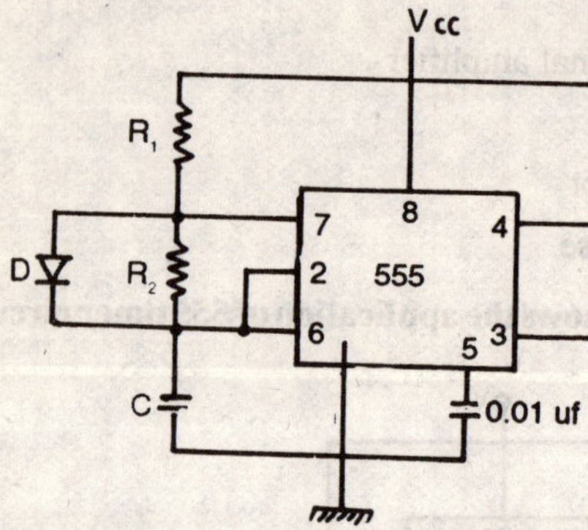

Fig. 15.17.

Q. 24. Recognize the circuit shown in Fig. 15.18.

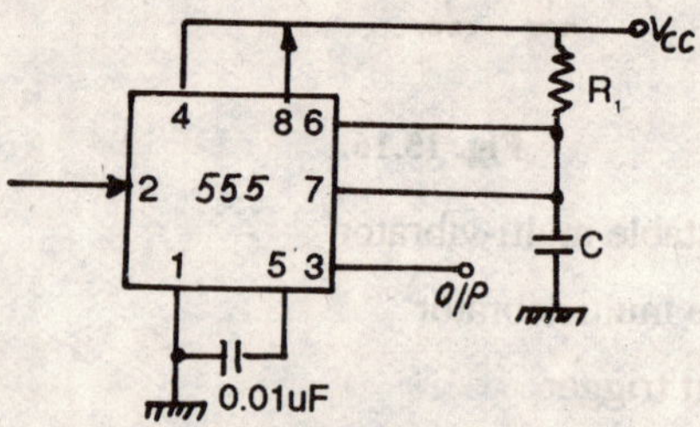

Fig. 15.18.

(A) It is an astable multivibrator

(B) It is a single shot.

(C) It is a bistable multivibrator.

(D) None of the these.

Q. 25. Recognize the circuit shown in Fig. 15.19.

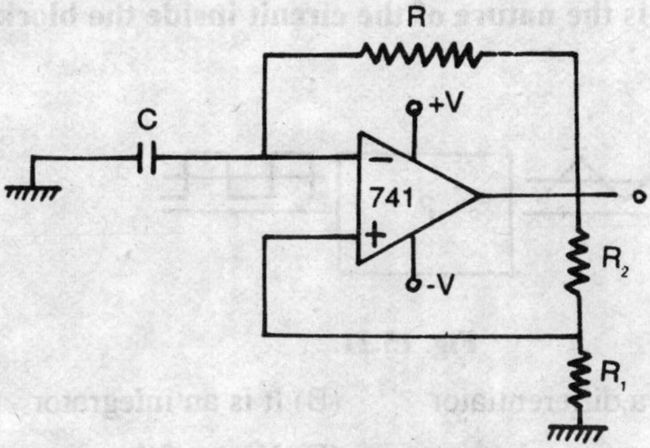

Fig. 15.19.

(A) It is a bistable multi-vibrator.

(B) It is a single shot.

(C) It is a free running multi-vibrator.

(D) None of these.

Q. 26. Recognize the circuit shown Fig. 15.20.

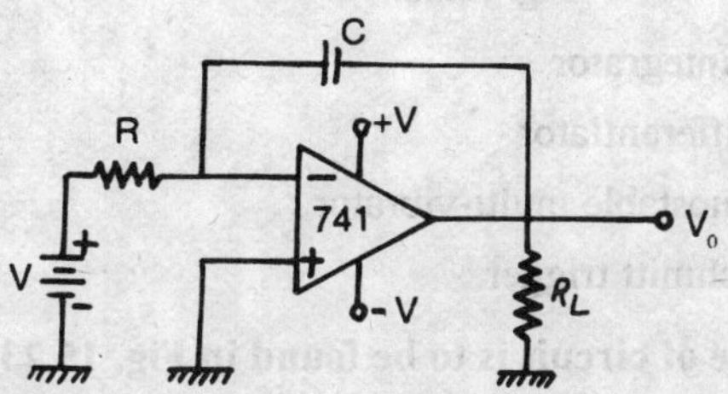

Fig. 15.20.

(A) Ramp generator

(B) Free running multi-vibrator

(C) Triangular waveform generator

(D) Monostable multi-vibrator.

Q. 27. What is the nature of the circuit inside the block in Fig. 15.21?

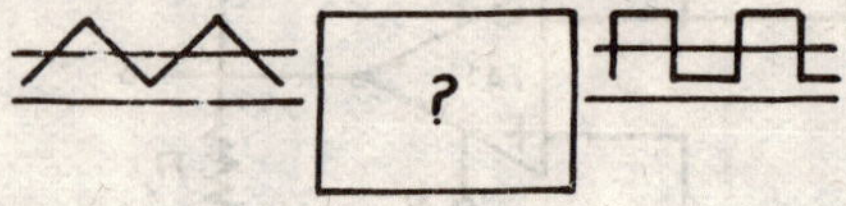

Fig. 15.21.

(A) It is a differentiator (B) It is an integrator

(C) It is a monoshot (D) None of these.

Q. 28. See the arrangement shown in. Fig. 15.22. What should be the nature of the circuit inside the block ?

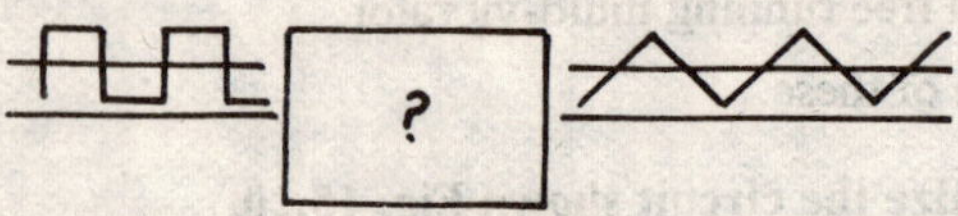

Fig. 15.22.

(A) It is an integrator

(B) It is a differentiator

(C) It is monostable multi-vibrator

(D) It is a schmitt trigger.

Q. 29. What type of circuit is to be found in Fig. 15.23?

(A) Monostable multi-vibrator

(B) Tank circuit

(C) Square wave generator

(D) Schmitt trigger

(E) None of these.

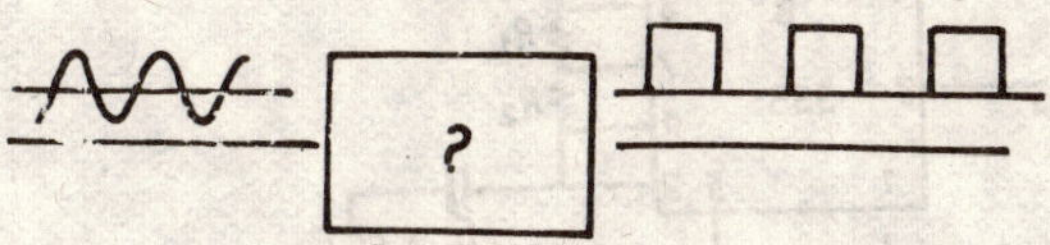

Fig. 15.23.

Q. 30. Recognize the circuit shown in Fig. 15.24.

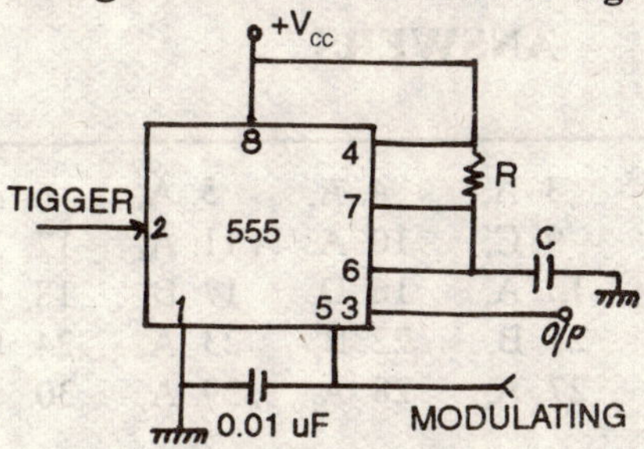

Fig. 15.24.

(A) Pulse Width modulator

(B) Pulse Position Modulator

(C) Monostable Multivibrator

(D) Astable Multivibrator.

Q. 31. Recognize the circuit shown in Fig. 15.25

(A) Pulse Width Modulator

(B) Pulse Position Modulator

(C) Monostable Multi-vibrator

(D) Astable Multi-vibrator.

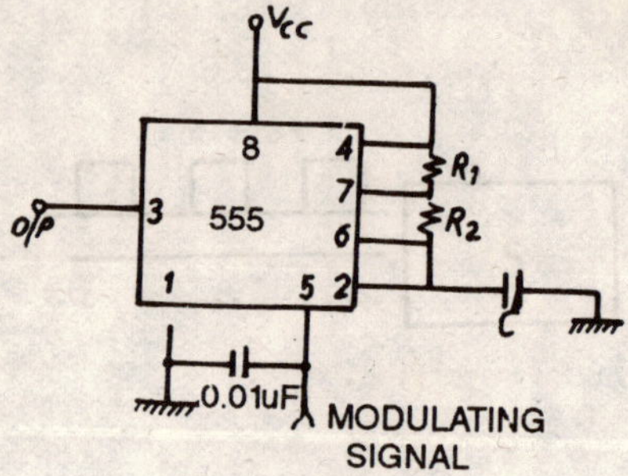

Fig. 15.25.

ANSWERS

1. C,	2. B,	3. A,	4. A,	5. A,	6. A,
7. A,	8. D,	9. C,	10. A,	11. A,	12. B,
13. A,	14. B,	15. A,	16. D,	17. B,	18. C,
19. A,	20. A,	21. B,	22. B,	23. A,	24. C,
25. C,	26. A,	27. A,	28. A,	29. A,	30. A,
31. B.					